KEEPERS of the STONE

Book Two:
Exile

Andrew Anzur Clement

One

The hallway was surprisingly narrow for a large dwelling. The floor was carpeted with rich, dark green fibers. The walls sported dark wood paneling between many doors, all of them shut. They were spaced at regular intervals on both sides.

It was also dim. The time was just after five in the afternoon – still rather light outside for this time of year. Yet, due to its location between two sets of rooms, the only illumination came from polished brass gaslights that hung as sconces on the paneled walls.

That was how the first floor hallway of the Pluckett family manor appeared to the man who now could call himself its head. Exhibiting strength of purpose that he did not feel, John Pluckett padded down the corridor, moving toward the room at its end.

He was peeved, to put it mildly. And he had been since the day the dark figure, which claimed to be one of the mythical Urumi, had come to demand payment of his debt to them. Whatever they were, he thought, they had promised him three things: the lordship of Yorkshire, a knightship of the Queen, and the restoration of his military career to its former ascendant trajectory. He'd gotten all three. For a time.

The major still had the first two. But, as it had turned out, the price that the dark figure demanded had required him to take a leave of absence from his duties as military advisor for the colonies to Her Majesty's court. John had not expected this eventuality. However, the figure had failed to return shortly after he had done its original bidding. That meant when he returned again to duty he would – the titles of Sir and Lord notwithstanding – find himself squarely at the bottom of the reassignment list.

Worse, John had needed to spend most of the last two months back at the manor. The dark figure under the bridge near Hampton Court had instructed him to keep watch over the charge it had commanded him to take and wait for its return. It was now early May; he was still waiting.

Sir Lord Major Pluckett neared the set of double doors at the end of the hallway. They were larger than any of the others and were inlaid with what appeared to be gold leaf in their beveling. In their centers, at roughly chest height, sat a carved, painted image of a white background out of which rose a red cross that contained five lions spaced at regular intervals. It was his family seal. As John Pluckett pushed the doors open, he allowed himself a small swelling of pride at the reminder of his place in British civilization.

Light burst forth from the doorway as Pluckett entered the large room that lay on the other side. The general decor was similar to that of the hallway, dominated by muted colors and dark wood paneling. Except here, much of the woodwork was more ornate, combining many different colors of lumber in ornate patterns. The high ceilings, also paneled in wood, contained about a dozen or so coffers. Three crystal chandeliers hung from the ones that ran along the center of the room. On the chamber's far side, high glass doors led to a balcony that looked out over the grounds of the estate. To the room's right end, a small but intricate chair sat on a slightly raised dais. This was the audience hall that the Lords of Yorkshire had used for centuries. It was John's favorite room in the house.

Now the hall was empty, of course. Sometimes, as he did now, the major came here to remind himself of his newfound standing. Yet, whenever he did so, Pluckett found that it also brought a reminder of how he had achieved that status and a knowledge, which he had come to reluctantly accept, that obtaining it involved a debt he was now obligated to pay.

Standing in the center of the audience hall, Pluckett grumbled out loud, to no one in particular:

"Not that it matters. This is my rightful place."

I just wish that I might be done with this foolishness, so that I can get back to my career. His mind finished the thought for him. The placement to the bottom of the assignment list, he told himself, was only a temporary setback.

Moving to one of the room's corners, the Lord Major pulled on a dark green fabric cord which hung from the ceiling in that location. A second later, he heard a distant chime, indicating that the manor's call system had relayed his request. He stood for a moment, inspecting his surroundings, until a section of the wall next to the hallway appeared to open. A servant, a young man with black hair in a navy-blue and white uniform, appeared from what was a disguised service entrance.

"Yes, Your Grace?"

"Another cognac, please," Pluckett rasped. Usually, he didn't imbibe before sundown. But there was simply nothing else to do, except wait and ponder what the dark figures – which he now had to admit held sway over his life – had in store.

The servant disappeared, leaving the brown-haired lord in solitude. Sometimes he wondered if this was it. If this was the end that the mysterious figures had in store for him. Did they simply wish for him to spend the rest of his days here waiting for further instructions about what to do with the charge that the Urumi had commanded him to bring here? No, he reasoned. That didn't make any sense. It just wasn't a viable long-term solution. They would return when….

Suddenly, he found that a tumbler about one-third filled with amber liquid was gently being placed into his right hand. Pluckett started slightly, realizing that he had become lost in thought. It was only the servant, having returned with his requested libation.

Without looking at the man, Pluckett asked:

"How is she?"

"You mean your friend's daughter, Your Grace?"

It was the lie that Pluckett had told the manor's staff: that the girl he had sequestered in a suite on the manor's third floor was the daughter of an acquaintance from the army who had come to stay with him while her father embarked on an unforeseen and urgent overseas tour of duty. The girl, he had told them, had many allergies and suffered from mental aliments. Thus, she was not allowed to leave her quarters.

"Yes."

"She is well, Sir."

"You have followed my instructions?"

"Yes, Sir. She has been given three meals a day and not allowed to leave her rooms."

"Have you spoken to her?"

"No, Your Grace, as per your instructions."

Pluckett had severely limited the number of servants who had access to his charge. And, he had forbidden them to converse with her, fearing that if they did so, they would begin to suspect the veracity of what he had told them.

"That will be all." John dismissed the servant and then proceeded out though the French doors that led to the balcony.

It was a pleasant enough view. John had seen to it that the grounds around the manor were immaculately manicured. Now that the weather had finally warmed, he found it pleasant to stand here in the afternoon, overlooking the estate that was finally his to control. Currently, many of its planting beds were bursting with yellow and blue flowers.

The Lord of Yorkshire sighed, taking a liberal swig of cognac. Yes, the view before him looked expansive. But, John also knew that he would not garner any further promotions in the military simply by standing here looking at it. He needed to get back. Pluckett thought again of the black figures whose demands had consigned him to his current presence:

If only that thing would come back and let me know what it plans. Then, I can finally be done with....

There was an audible rush of air behind him.

John turned to see a dark form standing in the comparative darkness of the doorway that led back into the audience chamber.

The head of the Pluckett household approached it.

"Finally. You return." He found himself snapping. "Why have you kept me waiting for such an amount of time? I have done what you asked. Let us finish this, so that I can finally be rid of you."

In response, the dark figure seemed to surge forward. John felt the force of an impact on his left cheek; the blow struck him to the ground. The glass of cognac went flying. Its crystal container shattered against one of the balcony's walls.

Quickly regaining his bearings, Pluckett struggled to his feet, a mix of shock, outrage and indignation crossing his slightly inebriated features.

"How dare you!" he roared. "I am the Sir Lord Major Pluckett! You will not treat me in this manner!"

Keepers of the Stone: Book Two

"All of which you owe to us, Mr. Pluckett. I will treat you as I please."

"You will treat me with respect," he emphasized the third word of that sentence, "or I will not consent to grant any further requests from your...."

Again, the figure moved in a flash. This time, John felt pressure applied to his neck as the Urumi whirled around to his back, catching him in a chokehold as it did so. The figure dragged him towards the ground. Pluckett was not exactly sure how long the figure held him like that, blocking his air supply. It was not until he began to see darkness encroach upon his vision that it relented. Moving in another blur, it turned to face him once again.

"You will consent to whatever I demand or I will kill you," the dark form's voice rumbled.

Still gasping for breath, Pluckett stared at the dark form for an entire minute before finally answering. His voice carried a slight edge, which could have signified either anger or fear.

"You would not dare. People will ask questions."

"They will ask. But, they will not find even a trace of my presence."

Another silence.

Eventually the Lord of Yorkshire responded in a low register.

"Very well. What do you request be done with the girl?"

The dark figure moved slightly; Pluckett feared that it was about to attack him again. However, a small piece of parchment merely fluttered down from the dark mass. The paper landed on the balcony's marble floor. Grudgingly Pluckett moved to pick it up, noticing the mark of the black serpent that graced its upper left corner. He took in the text written below:

Take your charge to the nearest port.
There, you both shall board the next available ship to....

Both of John's eyebrows shot up as he saw their intended destination. Of course, he reflected, exchanging the girl's unconscious body with the human-shaped facsimile that the Shadow Warrior had given him had been simple enough. Pluckett hadn't even needed to coax her from the schoolhouse, to which the dark form had directed him. She'd apparently wandered away from it in freezing weather – *rather stupidly,* the major thought – all on her own. This was different: after two months of waiting, the Order was ostensibly demanding that he travel with her to some obscure overseas destination. It was yet another setback.

"This is ridiculous!" he harangued the figure.

The dark form remained unmoved.

"Your understanding is not required, Mr. Pluckett. Only your cooperation."

There was a silence. It would have been an understatement to say that Pluckett disliked what he was being asked to do. Not only because it required him to leave England, the seat of where his power lay, but it also did not appear that the instructions offered any further information as to what to do once he had arrived with his charge. Instead of being asked to simply release the girl into the figure's clutches, or to kill her outright, these instructions simply ordered him to move her. Instead of a severance of his relationship with the mysterious dark form that now stood in front of him, the note John held in his hands amounted to a continuation of it.

The original terms of his agreement with the Urumi came to Pluckett's head. He decided to attempt pressing the figure on one aspect of their accord.

"We agreed that my additional services to you would only be required because I could not give my daughter directly into your service. Surely, you must have her by now."

A noise, sounding from the Urumi's deep register like a grunt, emanated from the dark form.

"Your daughter…." There was a slight pause. "Yes. She is not yet in our clutches. The time will come, perhaps sooner than she expects, when she will serve us. However, these extra obligations were required of you because you did not have her close at hand during the time of our agreement. Even when we take her, you are not absolved of those duties."

A beat.

"Very well." Pluckett grated. He had thought it worth the try. Over the past few months, he had come to the realization that what the Urumi had just told him was probably the case. He had not expected that attempting to clobber the figure over the head – assuming it had one – with its own contract would meet with success. Still less than enthused, John continued:

"What am I supposed to do once I get to this ... place?" There was a slight hesitation before he settled upon the final word, as if the Lord Major was unsure a more specific term would do it too much justice.

"You will wait," the figure said simply. Its tone made clear that this was not a polite request. All the same, it rekindled the Briton's ire.

"Wait!" The Lord Pluckett simmered with anger. "All I have done for the past couple of months is wait for your return. This is intolerable!"

"It is your choice, Mr. Pluckett. If you do as we say, you will wait. If you do not, you will die. It makes no difference to us."

John's face became even redder with anger. When he spoke, the man struggled to keep his voice level.

"How much longer will I be expected to wait?"

"We will give you more information once matters have become more clear. In the passing of roughly sixty days and nights."

"*Two months*!" You want me to go *there* for two months?" John yelled in disbelief. He paused for a moment, took in and then let out a sharp breath, attempting to calm himself before speaking again.

"Why must I go now? Why not stay here for two more months? I could still do whatever else you require at that time."

The figure let forth another deep sound. If it had been spoken in the normal register of human speech, it would have sounded like a sharp laugh or a bark. As it was, it came out more like guttural rumble. Still, its taunting purpose was clear.

"Because..., " the dark figure let out the sentence slowly, "the one you have been assigned to hold for us has escaped her quarters, just now. And, she had help."

Sir Lord Pluckett's eyes grew wide as he took in this information. The implications of the figure's statement were far from lost on him. If the girl managed to get away and told others of what John had done to her, he would find himself in prison – Lord of Yorkshire, or no. That was assuming that the figure standing in front of him did not decide to simply kill him in the event that his charge managed to abscond successfully.

Pluckett rushed past the dark figure, back into the audience hall. He headed straight for the fabric cord in what was, from his perspective, its far corner. John pulled it, hearing the reassuring sound of the chime.

Only moments ago he had wished for liberation, victorious freedom from the sway of the dark warriors. Now, he only hoped that it was not already too late to salvage his end of their agreement.

Two

The date was May 3rd; it was Nell's birthday. The dark blond-haired girl looked up at the calendar that hung on the wood-paneled wall of the room, which had become her prison over the last two months.

Her quarters were spacious. Well appointed. They included one room that comprised a sitting area, complete with a shelved wall filled with books. Opposite, a large picture window offered an expansive view of expertly manicured gardens. The next chamber constituted a spacious bedroom, complete with a mahogany canopied bed.

The same few servants came to deliver her meals three times a day. They said little. At first, all of them had eyed her with a certain condescending suspicion. It was as if, in a way, they feared her.

Now, Nell knew why.

At first, their manner had caused her to think – and to internalize – that there must be something wrong with her. For a while, she had almost come to believe it.

How could she not? The last thing the girl remembered before waking up in the quarters that she now inhabited was wandering away from the school, where her relatives had enrolled her since her unwitting arrival in England.

She had awoken in the bed that lay through the door in the sitting room's left wall. Nobody would talk to her. When Nell asked where she was, what had happened to her, her queries were met with a melancholically sympathetic silence. She had tried the only door that led in or out of her chambers. It was locked from the outside.

Nell continued asking every time someone entered her chambers. She did not know what exactly was going on. But, the girl suspected. She was no stranger to being kidnapped as a pawn in someone else's game. The Egypt-born girl knew from Stas's example that she had to keep trying no matter what; she was stronger than he knew. None of the servants answered her questions day after day. Still, she kept trying.

After a time, the two spacious chambers in which the daughter of an English engineer found herself became a cage. Sometimes, Nell pretended that she was a bird who dreamed of once again being set free.

Stas will come. He will save you. The thought had crossed her mind many times. It still did. Even though the days passed. It was as if that faith in him, that admiration, was what kept her going. And so she had read the books on the shelves – mostly forgettable English plays from the Middle Ages that she did not much care for. She had stared out of the windows and practically memorized the contents of a relatively recent encyclopedia and atlas, which she'd found on the shelves. Nell had asked where she was, why she was there, and what would happen to her every time someone entered her quarters. And, she had done so with all the tenaciousness of a terrier. Week after week. Month after month. It was to no avail.

Then, the day had come when it had not been. The young black-haired porter who typically served her lunch had come into Nell's sitting room. The girl had been in a foul mood. She was frustrated, understandably, with her circumstances. The man set the tray onto the coffee table in front of the couch where Nell sat – idly perusing a gruesome tale entitled *Doctor Faustus* – when she looked up at him. Her gaze pierced directly into his eyes. The sequestered youth asked a different question this time; one not directly related to her current predicament:

"Why are you doing this?"

"You need to eat," the man replied softly.

"I am not referring to my lunch."

The man – in her young eyes he was an adult – turned to go.

"Why are you keeping me here!" she yelled after him. "I want to go. Why can't I leave?"

The man turned slowly. "Because your father's friend says it would adversely affect your mental state."

Nell looked at the porter for a time, cocking her head. Again she took voice, as if both explaining something that she already knew, while coming to a deeper realization of it at the same time.

"My father has no friends that would do something like this." The young girl gave a pause, mental wheels turning. "You think I'm crazy." It was not a question. The black-haired man gave no reply. But, he continued to look at her. Shaking her head as if in confirmation, Nell continued:

"You think I'm crazy because I don't want to be here."

Nell began to pace around the room like a trapped animal.

"I've started to wonder myself if that's the case." Her tone was forlorn; despite herself, Stas's best friend felt her eyes begin to water.

The porter turned again to leave. Nell's voice stopped him. It was not that she yelled. Precisely the opposite. What she had to say was more for herself than for yet another waiter who served but did not obey her and, therefore, did not have to care.

"Stas. I know. You will come for me. You…." Nell stopped to take in deep breaths as she struggled not to break down completely. "You promised."

The servant who had customarily served the captive her noon meal flinched visibly. He turned, looking directly at her face. Nell's cheeks were moistened with tears. But, the girl still managed to tilt her head upwards to meet his gaze clearly. She did so, directing eyes that had seen more already than most Brits her age could even dream of, directly toward his. Though she cried silently, she exuded a desperate confidence.

"No one knows you're here. Except for your father." The black-haired man replied, not unkindly.

"If my father knew I was here, he would have come for me. But he never seems to know where I am. Stas. He always protects me…." Her voice trailed off as Nell struggled to stifle a sob.

The porter sighed. He gave her that same pitying, condescending look that most of the others who had entered her suite used to regard her.

"Your father will come back. Don't worry. His tour of duty should only last a few months."

Now, it was Nell who appraised the man. Her eyes burned with certainty even as moisture streamed from them.

"My father is an engineer. He isn't even in the military!" The girl's tone was clear, incredulous. Yet, at the same time, her voice shook slightly.

Apparently trying to console her, the porter stooped. His gaze met hers at eye level.

"It should be all right," the man began. "Lord Pluckett has told me of your father's wishes…."

Nell did not wait for him to finish.

"I don't know any Pluckett. I am sure that my father would never allow anything like this. When he sent me from Madras, when he sent me away from Stas – again – he said it was because he wanted me to be safe."

Nell continued to look the man directly in the eyes.

"When will Stas come?" The child of empire pleaded in a voice barely louder than a whisper. Despite her fear, the girl continued to hold the porter's gaze.

The servant stood up suddenly. He appeared to stare off into space, his black mop of hair dangling loosely around his head. He stared, as if not wanting to even look at the girl. He had been charged only with the simple task of serving lunch.

"He won't," the man almost whispered. "Nobody but our new Lord of Yorkshire knows that you are with us." He nodded as if finally understanding something. Then he turned to Nell, staring directly into her intense eyes.

"You're not crazy, are you?"

Nell did not immediately respond. Then, finally, she confirmed:

"Here? I fear that there are days when I wonder if I am not."

The young man with black hair nodded. He turned about and stepped back, seeming to repel away from her in a disorganized fashion until he hit the wall on the far side of the room next to the bookshelves. A tapestry hung upon it. He grabbed onto it as if attempting to regain his balance. In the process of doing so he revealed – with a clandestine point of his finger – that a keyhole rested in the paneling behind the fabric.

The porter's features suddenly hardened as he regained his balance.

"If you're just counting on some boy to get you out of here, then you are crazy." He had said the words loudly, harshly. But the black-haired man held her tear-streaked gaze until he exited though the door that led to the hallway.

The next day the same man had come to deliver her lunch. He had refused to talk to her, even as she asked him the same questions that she had so many times before. The black-haired porter had said nothing as he set the sterling tray containing her meal before her, bending down in order to do so.

Then, he allowed his face to pass near Nell's left ear.

"Teach yourself," he said.

After, the man stood up abruptly and took his leave.

Nell ate her meal mechanically. The lamb shank with mashed potatoes and carrots seemed to taste like cardboard. What could he have meant?

The daughter of an English engineer had been about to dismiss the porter's statement as a taunt as she finished eating, flinging the hunter-green napkin back onto the tray and settling back into the couch in a gesture of flippant annoyance. Something fell out of the napkin's folds onto the carpeting.

Nell's brow furrowed as she took stock of it. Then, she bent down to pick it up.

It was a paper clip. One of its tines was bent outward. The girl pondered the object for a moment before coming to a realization: this thing that the black-haired porter had given her must have something to do with the keyhole he had revealed the day before.

And so, armed with nothing but a paper-binding instrument, Nell had attempted to free herself, doggedly. For weeks she had awoken shortly after midnight and attempted to open the egress of her chamber's concealed entrance. At first, she had met with little success. The porter had continued to deliver her midday meal. He had offered no further signs of sympathy.

Now, it had been three nights since Nell had finally figured out how to move the lock's pins in just the right way so as to open the passage. However, she had no idea what lay beyond. She equivocated for those few days. At some point, she still hoped that Stas would come to rescue her. But, it seemed like the girl had sat in her suite for practically half as long as her adventures after being kidnapped in Africa. If what the porter had said was true, she reasoned, Stas could not know of her true whereabouts. That meant he was not on the way.

Nostalgically, Nell again recalled their journey through Africa. She had been eight then. With a sad irony, the girl recalled how much she had aspired to be as old as he was. Nell looked up to Stas as much now as she had then. She believed that he could save her from anything. Yet, if he did not know where she was, how could he seize the opportunity?

More than that, Nell remembered how much she had wanted to be as mature as Stas. Once, she had insisted that she was 'adult' enough to walk back to camp on her own. The Africa-born girl had attempted it, only to be confronted by a large black panther. Stas had killed it with his rifle before it could attack her. However, she had been struck by how he had been more unnerved by the experience of almost losing her than she had been at narrowly avoiding an apparent mauling. In the end, it was she who had consoled him.

Stas's friend had not understood it at the time. She wanted, and still did wish to be – in the abstract – like him. Independent. Confident. Valiant. Yet, she did depend on the Slav.

Now, Nell faced reality. Stas was not coming. She was on her own. This was independence. Unlike in the camp of the Thags, her sequestration in the manor had given her time to reflect upon her situation. The uncertain prospects scared the girl to her core.

Once, back in Egypt, Stas had boasted that he – one who was finishing his fourteenth year – should be capable of valiant deeds. Beyond the odd syntax with which he had spoken those words, Nell now found motivation in them. On this day, she was eleven. Only three years younger than Stas had been upon their kidnapping in the Sahara. As she prepared to attempt escape, Nell gathered solace from the following thought:

Stas used to tell me that he would save me because he was older. Well, I am now almost that age. If he could do those things, so can I.

Mustering her courage, Nell noticed that it was beginning to grow dark. Now was the best time, she thought. There was still enough activity to mask her escape. But, not so much that someone who recognized her was likely to notice her presence outside of this gilded cage. Silently, she moved to the tapestry. Pushing it aside, she took the bent paper clip from her nightgown. Pressing it into the keyhole, the girl pushed sharply up. The hidden door gave outward with a slight creak.

She moved forward, alone, into the darkness.

As soon as her eyes adjusted, Nell noticed that the corridor she found herself in was far less ornate than the environs to which she had grown accustomed over the past couple of months. Unshaded gas lamps lined bare brick walls. The sparseness did not faze her in the least, of course. However, it was the contrast that caused her to daze momentarily. For an instant, she considered darting back into the familiar safety of her quarters. But one of the most obstinate voices in her head made itself known:

If Stas can do these things, then I can as well. I will make myself capable.

Gathering her wits, Nell chose a direction and moved in it. The girl repeated the phrase to herself continually as she did so. It became a mantra.

Eventually, she found a staircase made of rude stone. It led downward and she headed in that direction. She descended three floors until the stairwell ended. Nell moved back in the direction she had come, noticing that it was beginning to get a great deal more crowded. A few of the servants cast curious looks in her direction. What was a girl in a nightgown doing in one of the manor's main service access corridors?

In response, Nell kept her expression carefully neutral. She made a conscious effort to maintain a mien that betrayed no hint of uncertainly, as if she had the right to belong exactly where she was. On the inside, however, the girl's mind was a tumult of anxiety and fear. She could not honestly say how well she was able to hide the fact. That only heightened her state of worry.

Suddenly, Nell noticed a shift in the attitude of those around her. Many began to whisper, and suspicious glances turned in her direction. One of the serving staff – a man she had never seen before – began to dart towards her.

Nell ran. She took the first right and found herself in the scullery. At least ten individuals occupied it, each of them sporting soiled servants' garments. Nell could sense that a crowd of at least three or four was pursuing her into the manor's kitchens. At the same time, the girl noticed that a middle-aged woman had somehow appeared not to notice the commotion. Wisps of gray in her hair, she stood chopping carrots on a large table of butcher's block.

The girl dashed towards her. Sticking her foot forward she ground down on the woman's instep, causing her to yelp and drop her chef's knife. Grabbing it from the table, Nell turned and backed towards what looked to be the scullery's outer door, facing her pursuers.

Nell was uncertain as to what she should do. For the first time in her existence, the power of life and death was in her hands. She had no desire to throw the knife at her pursuers. She had seen Stas kill her kidnappers after they had made it into the heart of Sudan. As his closest friend, she knew internally how much it had cost him. On the other hand, she had seen her kidnappers in India kill in cold blood; she did not want to be like them. Besides, the thought crossed her mind, she had no idea how to effectively throw her own weapon to deadly effect. Nell made up her mind.

She continued to back towards the door, holding off the servants who had followed her from the hall. She could hear the kitchen staff closing in behind her. Obviously, she could not fight them all. It was time to make use of her only remaining advantage.

Suddenly, Nell dropped the knife and bolted for the door. The servants moved after her in a wave. But, the Egypt-born girl managed to use her small size to evade them all. She reached the door and pushed upon it. The unvarnished paneling gave, allowing the escapee into the refreshingly brisk air of freedom.

The girl had not gotten two steps beyond it when a sudden force knocked her down and to the side. She looked up to see a dark form, flanked on its right by a formally dressed middle-aged man with brown hair and bleary eyes. There was nowhere left to go. Nell sat next to the exterior brick wall of the manor, subdued. She stared up as the black figure moved again.

It rushed toward her, colliding with her head. Nell felt her own skull make a cracking sound against the building's masonry.

Then, all was silence.

Three

They descended the foothills of the Rocky Mountains, although none of the trio thought of the range in those words. A wide valley spread out before them. Malka knew what was coming. She did not like it.

Their passage through the mountains on horseback had been without incident, though chilling. Literally. While the spring thaw had brought a rise in temperatures, it did not mean that the group had escaped the chill of snow banks, which had yet to melt. Despite the apparent warmth of the sun, the Thag had bitten her inflamed fingers repeatedly during their passage.

Their native traveling companion had remained sparse with his words since the beginning of their sojourn together. The white-haired man had instructed the Sect's leader in the arts of throwing the ax. But, beyond that, he had stayed rather tight-lipped since their meeting.

For her part, Liza – aside from her general air of annoyance – regarded the man with discreet suspicion.

And yet, Malka had spoken to him during her training sessions with the triangular blade. The blue-eyed girl told him of some aspects regarding her past: how she had lost the only family that she had ever known in what she now feared was a fool's errand. How the continuation of their memory was one of the few things that kept her going. And, how they never had truly accepted her, despite her loyalty, almost until their end. She did not reveal what had happened on the day she had finally faltered. That was too much for her to contemplate.

At each revelation, the disheveled native had simply nodded, as if in a sad gesture of understanding or sympathy. But, he had yet to volunteer any information regarding his own circumstances. And although Malka had pressed on more than one occasion regarding the status of Liza's 'kind,' the man had remained succinctly tight-lipped. Instead of answering the Thag's queries, the long-haired native would often stare into the horizon, as if in search of some answer himself.

"When she is ready." That was all he would say.

Malka's training with the ax had gone smoothly. Husain's protégée could now proficiently throw the weapon so that its sharpened end reliably met with her intended target in practice exercises.

Now, as they descended into warmer climes, the same mysterious man who had taught Malka the newest skill in her offensive arsenal brought his mount to a halt. He held up his hand, indicating that the two who followed behind him should follow suit. Then he pointed to a location near the horizon.

Malka could make out a glimmering line. Reflecting the sunlight, it moved at an oblique angle to her current orientation. Periodically, settlements cropped up around the line of metal. The Thag noticed that the Indian was pointing directly to one of the smallest ones.

"You must go there," he intoned. As usual, the man's voice was unintelligible as to whether he regarded his statement positively or negatively.

"Very well," Malka replied with a small melancholy smile. It was an acceptance of a man she did not truly know, but understood somehow that she could trust.

"Why?" Liza growled, her voice dripping with suspicion. "We are fugitives, as I'm sure you now know. Head into any kind of local settlement, no matter how small? And bam!" The felinoid slapped her hands together. "They throw us in the slammer. So, yeah. Guessing that's not the best idea."

In response, the man nodded sadly. He kept his tired gaze fixed on Malka, the yellowed whites of his eyes set in acceptance of an inescapable fact as he did so.

"Your companion is right. You have told me of how you have gained the paper which you carry…."

Here the felinoid interrupted him:

"Which, as I have groused, was exactly not the best idea either, considering…."

The Indian continued with no hint of malice, as if the black-haired youngish woman had not spoken. He did not mean such an action as a slight. He continued with an accepting equanimity.

"We are still too close to the settlement that you have fled from. They may be able to trace any bills that you spend. You will have to work for passage on a train to New York."

"Very well." The Thag's response was immediate.

"You may not know what that entails. Due to the color of your skin, they will regard you with contempt." The latter sentence was spat with a tone bordering on hatred.

Malka shrugged. She dug her heels into her horse and began to move forward. Scowling, Liza did the same.

The dark, weatherworn man did not move.

"The two of you must go alone," he reminded them.

Malka stopped, turning to look back towards the man who had become her guide over the past couple of months.

"But why? I've no idea of what lies ahead." The Thag kept her features neutral. But there was a desperate, pleading look in her sky-blue irises. Both Liza and the Indian took note of it.

There was a short silence. The man sighed, eventually deciding that it was time to divulge something. His eyes darted back and forth between Malka and the felinoid as if he felt strongly that both should be included in the conversation.

"As I have told you, this is the land to which my ancestors belonged. When those who run the mechanical beasts...," he moved his head to indicate the steel lines toward which Malka and Liza were to direct themselves. "When they took it from us...," again another pause as if the disheveled man struggled with the reality of recollection, "I was one of the few to escape. I fought them however I could. I became a fugitive – a criminal. Why do you think I found you on the other end of the mountain range? I am not welcome here."

"And yet you returned?" Malka asked the question. Liza simmered silently in the background, as had become her habit.

The older man nodded slowly. Knowingly.

"Yes. Meeting you has reminded me of the nobility of that fight – no matter how futile. My return will probably mean death, but it will also mean that my people will go down with struggle." He offered a brief, wry smile. "That contest cannot be best served by simply walking into one of their strongholds."

The Indian, who ordinarily avoided making eye contact, looked up. He stared directly into – and beyond – Malka's eyes.

"I know something of who you are," he continued. At this, Malka furrowed her brow. "Struggle, though futile, can be more important than existence."

The Thag dismounted from her horse and approached the morose man. There was so much that she wanted to ask him. Of course, the girl had told him something of her background since they had begun their combined journey, but nothing of the object she carried. How much could he know of her struggles?

On the other hand, the white-haired Indian had seemed to display a certain kinship with Malka's situation. He had divulged nothing of his own past. If the Thag did not have secrets of her own to keep, she would have been suspicious of him. Yet, it was precisely because of those secrets that the blue-eyed girl felt a certain solidarity with her guide. She did not truly know him, but wanted to.

Finally, looking up into his bleary, intense eyes, Malka summed up all of the inquisitiveness that welled within her into the following, tentative query:

"Who are you, really?"

The man nodded slowly, his unkempt locks moving about his head. He offered a wan smile that betrayed wisdom, not unlike the smile Husain had always given her shortly before stating a truth, which was in fact a lesson.

"My name is not important, Malka. It was given me by my tribe. They, as I knew them, are gone. As yours has perished." Again, he regarded the Thag. "I have done what I needed to do."

Without warning, the Indian removed the ax from the waist of his trousers. He flicked it into the air in an arc over his head, causing the handled blade to spin. Malka's guide caught the weapon by its metal blade's oblique edges as it returned towards him. He extended it, handle outstretched, to the Thag.

Her blue eyes widened.

"No, I could not possibly….," she whispered. "You require it for your own quests. More so than I. How will you assail your targets?"

The Indian smiled briefly.

"My fate will not be long. I will fight them until the end in my own way. You have reminded me of the importance in doing such things. Against their firearms, such a blade cannot hope to prove itself with effect. In this way, it carries more meaning."

The man dismounted from his horse. The ax handle was still outstretched to Malka.

"Take it. It is a gift from my people to yours."

"But…," the Thag's appointed leader replied, "I have no people."

The man smiled sadly. He bent down, staring only a few inches from the Thag's gaze while pressing the blade into her left hand. "No longer do I."

Malka took in a sharp breath. Even in the months through which she had known this man, he had revealed little concrete information about his past. Yet, from what he had told her now, she felt an instinctive kinship with his situation.

The leader of the cult of Shakti slowly closed her fingers around the handle, accepting the weapon. Her eyes began to water. Then she said: "You can keep the horses."

Without offering any hint of emotion, the man stood.

"I thank you." He turned and began unloading the bags full of cash from his own horse as Malka did the same.

As he placed the canvas bags before her, the protégée of the Sect's leader noticed that Liza had elected to dismount as well. However, it was clear from her manner that the felinoid did not intend to aid in the transfer of materiel.

"Malka! How the hell do you really expect to show up at some random filling station with bags full of cash and get a job running trains? It's going to attract a load of suspicion."

Again, the Thag shrugged.

"I am a Thag. I will figure something out."

"Really? Amazing. That's an identity and a promise. But, for your information, it is *not* a plan."

For the moment, Malka ignored the felinoid. She directed her glimpse toward the Indian man. Liza threw up her hands and looked toward the sky as if asking for guidance in regards to the situation.

"May the quest for your own people go well," Shakti's follower imparted in a tone of reverence.

"And yours. Use my people's blade wisely. Employ it. Protect the most sacred charge of your own."

Malka could not help but ask, "How do you know of my charge?"

"Yeah, really," Liza echoed, though more out of blatant suspicion than wonder.

The Indian did not reply. It was not as if his intent was to be mysterious or obfuscatory. Instead, he felt that further explanation was not necessary. The Indian moved to tie the horses together. Then he headed them northeast and moved off, leaving Malka and Liza standing there with three saddlebags full of cash. Malka looked after him as his white hair disappeared down the slope.

"Great. Now what?" It was Liza who spoke first, placing her hands in front of her midsection as a mild accentuation of her query.

Malka moved to pick up two of the saddlebags, throwing them over her shoulder in response.

"Now, we move toward that silver line."

"You heard that man. If you try to use any of this money from your and Henry's escapade to buy a train ticket, they'll suspect us."

"You heard him as well. I can get a job. Use that for passage."

"A job? Malka, you don't know the first thing about trains. Why the hell would you think that they would do anything other than send us packing?"

"What I do not know I can learn."

"You can learn?" Liza was incredulous. "This, from a girl who has been taught her whole life that she can be nothing but a thief?"

Malka turned to glower at the felinoid. "My Sect is gone, Liza; they are gone because of me." Here, the last of the Thags turned about. She gazed in the direction that the Indian had gone.

"Do not accuse me of forgetting their memory. But I must continue as best I can. Besides, it is merely a means to an end."

"Really, and does that end include traveling across half a continent to rescue some boy who you took into the fold for no good reason in the first place?"

"As I said. My Sect is gone," Malka replied with a sigh. She began to move further southeast into the valley.

"Fine. At least it's a win that guy didn't try killing us for the cash over the past two months." The felinoid groused as she picked up the remaining saddlebag. Then the milk-skinned girl followed her charge into the plain below.

Four

Malka approached the passenger train. Its engine sat under a large spigot, from which water gushed out of a tank into its black hull. Near it, a few young men in wrinkled railway uniforms, some of them soiled with soot, loitered about in boredom. It was as if they were annoyed by the necessity of the stop, but not particularly eager to continue with the task of facilitating the conveyance's journey, either.

It was the one facing away from the train's engine that first noticed the approach of the Thag. He fixed his brown eyes on her, such that the other five or so young men immediately turned their craniums to take note of what had distracted their ringleader's attention.

The sight they perceived was peculiar to behold. A darker-skinned young woman with surprisingly light hair and blue eyes visibly labored under the weight of three saddlebags. A few paces behind her, a black cat followed. The sight only aroused their suspicion.

"Hey you, Squaw. What brings you here?" said the first one to see her. His uniform was particularly soiled.

"Where did you get those bags? You don't got any horses," a second, positioned to his right, leered. His garments were in slightly better shape.

As if in answer, Malka kept moving toward them. The girl displayed no outward sign of reaction.

"She's an Indian. They aren't even hers. It's obvious she stole them," the first replied.

The Thag kept moving forward.

"You're right. Good for nothing. What are we gonna do about it?"

"Let's see what she's got," replied the first. His tone indicated that he regarded his booster's question as rhetorical.

Now, Malka was practically in the midst of the group. She stood, ready to make her case, or fight if necessary.

The band of young men began to close around her. As they did so, Malka's hand went to her left boot, instinctively moving for the dagger that lay concealed there. But then the girl realized that her mobility would be severely limited, due to her payload. Despite what the long-haired, indigenous man had told her, the blue-eyed disciple of Husain had expected to at least be given a hearing. In the camp of the Thags, she had been taught how travelers were supposed to respond. Here, however, that was clearly not the case.

Malka stood her ground. The men closed around her. She resolved to fight as best she could. Even if that would place the safety of what she carried in jeopardy. Such was the way of her Sect. Still, even as the subcontinental native did so, she had to acknowledge that her chances of holding off six others simultaneously were slim, especially while her burdens restricted her movements.

It wasn't that Liza could do much to help her, either. The felinoid had changed into her catlike form shortly after the two had begun their descent into the valley. Handing her saddlebag to Malka, the black-haired young woman had claimed it would attract even more suspicion if she were to appear in her human form alongside the Thag. The blue-eyed girl was still not sure if Liza's justification had been genuine, or just an excuse to get out of carrying a third of the ill-gotten cash. But, in any event, it would not do for an entire train full of passengers to see a domestic feline transform into a biped.

And so, the Thag reached for her knife with her left hand. With her right, she moved to retrieve the ax head that now rested in her waist belt. She prepared for a futile fight.

"What is going on there?" an older man's voice yelled. Sporting a beard, he was dressed in formal attire of black and white. A golden watch chain hung from the vest, which he wore beneath his tailed coat. His left hand carried an ornate walking stick of brass and dark wood as he descended from what appeared to be a private railcar, situated directly behind the train engine.

"Um, sir!" the first of the two replied as if surprised by the man's interest. "Well, sir, we were waiting for the engine to take on some water. Then this squaw showed up and we were keeping her from doing harm to your engine. We want to be good employees, sir."

"Tell me this: what harm, exactly, is one girl going to do to this train?" The man raised the knob of his walking stick towards the ringleader of the group of boys.

"Sir, perhaps you didn't see what happened...."

"I saw exactly what happened, all of it. Are you boys so stupid that you think I do not watch what happens on my railroad right outside of my window? First, I was none too enthused that you chose to take the opportunity of this filling stop to loiter outside of the locomotive rather than find some work assigned you by the company. But then I see you attempt to steal...."

"But, sir," the same youth replied, "she's an Indian. Can't be trusted. How do you think she came by whatever she has in those bags? She's up to no good. Why else would one of her kind come here?"

The older man harrumphed.

"As for your first question, it is none of my business. It is none of yours, either. As for your second...," the older man turned to regard Malka, "why *are* you here?" he demanded.

"I need a job," Malka replied simply.

The tycoon's forehead creased in suspicion.

"Why?"

"Because I need to get to New York."

Behind the Thag, Liza hissed in annoyance. At the same time, the older man's features betrayed surprise.

"Why do you want to do so while working for the Southern Pacific Company?" he clarified, as if assuming that the darker-skinned girl must not have understood the question correctly.

"I do not, particularly. Actually, I would rather not say. But I do need to get to New York."

Liza meowed.

"Give me one good reason why I should hire you, then."

"Because I need to get to New York City within two weeks. I am willing to work for passage."

At this the man appeared visibly annoyed, as if he was having his time wasted.

"All right then, answer me two questions: Have you ever worked trains before? And, why New York?"

"No, but I can do any job. Learn how to do so. Henry's in New York."

"Who's that?"

"He's...," Malka hesitated as she thought of how best to describe the boy she had rescued. "He is my brother."

The older man regarded her with a glare. "We are no charity service," he said, raising his head back toward the carriage he had apparently been traveling in.

"Henry! Get out here!" he yelled. Then he addressed Malka. His tone indicated neither belief in her contention, nor respect. "You're an Indian. Your brother's name is Henry? Not Crapping Cat?"

Liza hissed, at which the rail mogul let out a bark. "Well, my nephew's name is Henry, too. I'll leave him to deal with you. Maybe he can make himself useful, for a change," the man spat, then turned to walk back towards his private railcar. As he climbed in, he turned his head back to Malka and the group of men in front of which the Thag now stood.

"Oh," the man said the word as if he had forgotten something, while pointing to the young man who had first seen Malka. "You. You're fired."

"Fired? But sir! This is the middle of nowhere. I'm not due to be paid until the end of the month. How am I supposed to get home?"

"That's not my problem. Try to catch some rabbits as they run across the tracks for all I care," the older man grumbled as he struggled up the steep stairs back into his car.

A few seconds later, a younger, clean-shaven man disembarked from the same car. He approached Malka and looked her directly in the eye. Behind the Thag, Liza hissed at him as she had his older counterpart. The younger man flinched upon notice of the cat. Then, he continued.

"You will have to excuse my uncle," the beardless one told her. He was no less formally dressed, but his mien was much kinder.

"You want a job?" he asked.

"Yes. I need to save someone. He is like an ... adopted brother to me. Please ask no more."

The younger man regarded her for a time.

"My uncle has already ordered you to be sent away."

At this, Malka stared at the ground. Then, the younger man spoke again, a smile slowly spreading across his face.

"But he also just fired our main coal stoker. You'll be his replacement."

In response, Malka clasped the palms of her hands against each other and bowed deeply. It was the only way she knew of showing respect.

He motioned to the bags the girl carried upon her shoulders.

"You can stow these in the caboose."

Immediately, two of the young men, who had stood next to their newly fired ringleader, moved to collect the saddlebags. They picked them up, moving toward the rear of the train. One of them – the same who had egged on the group's ringleader earlier – stuck his hand into one of the satchels. His eyes grew wide as he did so.

The formally dressed man saw the gesture. He did not take kindly to it.

"Hey! You there! Thief. I see what you are doing."

The chastised worker in question tried to play innocent.

"Doing what, sir?"

"I saw you sticking your hands into a fellow employee's belongings."

"But, sir, she has...."

"I do not care what she has in there. Have you no respect for the concept of private property? You can join your friend; your services will no longer be required here as well."

The young tycoon turned back to Malka.

"I will store your belongings in my personal safe. My uncle and I will be traveling on to New York, before he moves on to his shipyards in Virginia. However, this train will only take us as far as Council Bluffs."

The clean-shaven man took out a small piece of paper from his coat pocket and wrote something on the back of it.

"Here, take my card. Show this to the conductor of the service we will be proceeding on. Assuming you prove yourself, with my name he will give you the equivalent position."

Malka took the piece of parchment. She looked at the characters, which were embossed with black ink.

"I thank you, Mister Henry Huntington," the Thag intoned. Then, without hesitation, she started into the engine compartment, while the suit-dressed man moved Malka's saddlebags into his rail car.

After he had done so, a black flash followed the Thag into the place of her new vocation. Shortly thereafter, the train departed. Two young men were left in the middle of an empty plain. Their eyes widened as the conveyance moved off into the distance. A white-haired man, with a train of horses tied behind him, began to circle them. With a cry of rage, the dark-skinned man charged on his mount, his long hair flying. The two attempted to run but they were quickly overtaken by the horses. They felt the pain of hoofbeats driving into their skulls. After, they knew no more.

***

Inside the engine compartment, Malka regarded the large black boiler that was meant to power the steam engine. Immediately following her entry into that space, an older man had handed her a shovel and told her to deposit a black substance into the large furnace object that sat at the front of her new workplace. Without hesitation, the Thag did as told. As she began to do so, the man pointed to a dial on the cast-iron contraption.

"Stop when this dial reaches forty. Then turn this lever twice left. That will flatten out our acceleration."

Malka nodded once. The girl continued with her task.

"Very well," the older man continued. "I must see to the passengers' tickets."

The disciple of Shakti ignored the man's last statement. She was here to work for an end, not to make friends or get to know an organization or those who worked for it.

However, it was only then that she noticed: two of the remaining four who had formed part of her 'reception committee' were present in the car behind her. They must have entered directly after her and Liza, who now crouched inconspicuously in a corner. Both advanced towards her.

"You just got Harry and Charlie fired. Those guys were our friends."

The Thag turned, instinctively brandishing the shovel as a weapon by turning it in front of her, as one would a quarterstaff. She passed it between her hands.

"Tell me why I should care." It was not a question.

The two kept their distance.

"So, we are done here," the Thag confirmed.

None of the two moved.

Malka nodded once, succinctly.

"Very well. Do you not have something to do elsewhere on this machine?"

Slowly, they both nodded.

"Brilliant. Then do so."

The two turned and retreated in the direction that the older conductor had gone. Malka ceased regarding the aperture. Using her shovel, she continued stoking the train's engine. The girl choked on black dust.

As the Thag labored, Liza sauntered past. She turned her green eyes to stare at the newfound rail employee. Malka met the felinoid's gaze.

"You're not even going to help, are you?" The blue-eyed girl sighed.

In response, Liza continued to stare at Malka. She jumped onto the warm boiler and curled her paws under her compact body.

Malka's protector looked pointedly at her charge. Then, she shut her eyes and went to sleep.

Five

The apex of the stone building that stood opposite the forested ravine was just barely visible. It had been that way for as long as Henry had been kept prisoner here. The black form had come and gone. It had made sure that he had been given sufficient nourishment. At various intervals, it would release the boy from the cast-iron trappings to which he now sat bound. Then, it would demand that he do calisthenics as it blocked the only egress from the ruined building's courtyard, even though it was already covered by a still-sturdy wooden and iron gate.

The son of miners had spent months in that manner, since his plan to escape into the ether with wads full of cash had blown up in his face. One moment the brown-haired youth's escape had been going well. Then he'd found himself in the clutches of a mysterious dark figure – probably a member of what his first kidnapper had called the Urumi.

At least Malka was human, the boy thought with a heavy sigh.

He'd had plenty of time to think. Sitting bound in the open-air courtyard of this ruined building, Henry had no idea where he was. But when he'd first appeared there, in the clutches of the dark figure, it had been much colder than on the plain from which he had been taken in Nevada. What was more, the topography and foliage were completely different – denser, coniferous. The ruined building in which he sat was composed of a dull gray stone unfamiliar to him.

As the American-born youth sat shackled to the ruins' cast-iron braces, he thought about how the dark figure rarely left the courtyard itself. At times, it would stare out over the horizon, in the direction of the other building across the ravine that he could not see in full. At others, the dark form would sit across the open space, seeming to regard Henry, though as far as he could tell it had no eyes. On periodic instances, the blue-eyed boy's captor would ensure that he was firmly secured to the building's wall and leave through the gate on the other end of the courtyard. It would usually return shortly thereafter with the carcass of an animal, which it would cook over a fire and then extend for him to consume.

At least, it was finally warming up. This lackadaisical routine had gone on since Malka's former captive had been brought here. At first, it had been much colder; Henry had been exposed to the elements. More than once it had snowed upon him. Then, it had rained. Often. It was surprising that he had not caught his death of cold.

More than that, the dark figure had never spoken to him. For the least part, Malka, and even Liza, had done that. As he'd sat chained to the wall of a former dwelling that he could not recognize, Henry had long been cognizant of an existential wish: he wanted to go home.

But, as he sat in yet another stretch of tethered solitude, Henry realized that he did not know where his home was. Home had been in his parents' camp, where he had always been the black sheep anyway. The camp was a thing of the past and his parents were dead. Henry now fully realized that if he were to be set free, he had no idea where he would go.

The boy had always been told that he was from America. But, he had little concept of what the reality of that place was like beyond the mining camp where he had been raised. It was not as if he had particularly liked that locale either, he realized. At some point, most there ridiculed him for having any interest beyond digging in the dirt for silver. That included, most strongly, those of his own age. What was more, upon entering Elko, Henry had feared that he would not know how to conduct himself in such a place. He had not liked the few he met there and had no trouble relieving them of their money.

The native of the camp may have been told that the US was his country, but he did not feel a part of it. In fact, since leaving his parents' now-defunct camp, he had felt even more threatened and out of his element.

Malka had been the only constant. At first, the blue-eyed boy had not wanted to admit it to himself. She had taken him into her troop against his will. The mysterious subcontinental girl with strangely light hair and blue eyes had saved his life – that was true. Yet, she had insisted on drawing him into her own thrall. Resultantly, he had drawn upon his own knowledge to hatch a plan for freedom, after his initial attempt had failed. It had almost worked that second time.

Henry leaned back against the cold stone wall, his hands bound near his left shoulder. He let out a deep breath. Sitting there, the young man realized that over the course of their sojourn together, he had gotten to know Malka and the strange, irate were-cat who called herself Liza. He had dismissed them as common thieves at first. But, he could not explain the black-haired girl's transformative abilities. Nor, despite his wide reading, could he say what held him captive now. As the native of the mining camp sat in an abandoned courtyard, he had to admit to himself that some of the stories he'd heard his captors relate might be true.

Further, he realized that he missed Malka and even, in a way, her curmudgeonly protector. True, he had not exactly asked to become a part of their band. Indeed, he had fought for escape in the best way that he knew how. Yet, they had faced, albeit unwittingly, life and death together. After arrival at his current place of incarceration, the boy realized that he had begun to feel a certain kinship with them. And he understood that even if he could escape now, he had nowhere to go but back to the girl who had first kidnapped him. Her quest was the closest to any mission – any sense of belonging – that he had left.

A sound jolted Henry from his reverie. He started.

The black shadow moved through the entry gate of the dwelling's courtyard, shutting the still-functional defensive aperture behind. It lugged the dead figure of a buck into the courtyard's center. The being started a fire, used its elongated flexible blade to cut a hind leg from the animal, and then held it over the flames.

It waited, unmoving, as the flesh cooked. Then – as Henry knew it would – the form held out the shank to the boy with one seeming appendage while untying his hands with the other.

"Um. Deer meat? Again?" the boy sighed. The figure remained unmoved. The brown-haired youth had not really expected a response. He had gotten none before. The shank continued outstretched, seeming to hover on the edges of the dark form's vague figure.

Eventually, Henry took it.

"Okay. Thanks."

The boy had begun his meal morosely when he noticed that, without warning, a second black figure had joined them in the courtyard. The two forms began to vocalize. Yet, it was in a tongue of which Henry had no knowledge. The boy stopped eating as he attempted to glean whatever he could from the conversation taking place before him.

***

Bozhena knew who it was that had entered her courtyard only shortly after she perceived his presence. This was it. He had found her before she was able to put her plan into action. At this point, she was sure her plan was better than anything that the Urumi's commander was capable of devising.

The Slav, still in her dark form, waited for the other to make the first pronouncement.

"Bozhena Alexevna." The deep tone of the other dark figure boomed.

"Ziya bin Mohammed Ahmad."

The dark form of the Chosen moved as if to circle her own.

"We are both our fathers' children, are we not?"

Bozhena knew that her commander had meant the question in a rhetorical manner. However, she could not help but offer a response:

"In a biological sense, that is self-evident."

"I mean our constitutions reflect our lineage." The Chosen's voice boomed. It was as if he had interpreted her response as a true lack of comprehension. Remembering her mother, she also took note that Ziya seemed to have little regard for the matrilineal aspect of his ancestry. It chilled her.

The Chosen's black form continued.

"My father declared war on the entire world. I have no doubt that my younger brother will continue to fight in that cause for my people. I myself am leader of those who serve our Dark Prince.

"And what have you done?"

The Chosen's inquisition seemed to burn in Bozhena's ears. It was not because his words rang true in her mind. Instead, it was because –

in spite of her own wishes – she was executing a plan which she judged far more likely to meet with success than whatever her leader had held in mind, since the flight of the Fragment's holder from the subcontinent. Focusing on him, the Slav framed her reply.

"My Chosen, forgive my absence from our sacred grotto. I have been in the process of executing a plan of my own. One that will see the Fragment delivered into our presence. And this youth," the blond-haired girl gestured to Henry, "consigned to the service of our Dark Prince."

The dark form advanced toward her, threateningly.

"What have you done to advance my orders? Where are my lieutenants? What is the status of the power-hungry one who I commanded you to develop? You do not report back, and then I must come and find you? Here?"

The one who was, in her true form, a blue-eyed Slav, bristled inwardly. The Chosen's refusal to explain his orders, combined with his lack of track record, had done little to inspire her confidence. At the same time, the girl remained convinced that using the Anglo-Saxon gentleman was an unrelated, or at least inefficient, way of obtaining the Fragment from its current holders. She said none of this to Ziya.

"Your lieutenants are dead," the warrioress replied simply. "Those who carry the Fragment killed them."

"You allowed this to happen." Ziya continued closing the distance, so that their forms were almost touching.

"No, my Chosen. It simply did." Bozhena did not back away from her Order's leader. She also did not point out that it had been on his command that she and those aides had pursued such a course of action. Instead, the girl chose her words carefully: "I have remained occupied. Executing a plan that will redeem myself in His eyes and those of your augustness.

"The one who holds the Fragment is troubled. For that, she bears a strange attachment to this one." Bozhena's form moved to indicate Henry again. "I will use phantoms of her past and the promise of his release to sway her."

"If it does not?" the Chosen's voice bellowed, unconvinced.

"Then, I will take it by force. And the boy will either die or become one of us. In any case, the promise of his release will bring the Fragment's keeper to me."

Ziya's dark form regarded her for a moment. Then, he intoned:

"See that it does. Yet, none of your justifications excuse delinquency from my orders. I expected that you'd be available to work our asset. Instead, I find that you have been here all of this time?"

The Sudanese Arab's ire rose as he pronounced the words. As he said them, he moved, striking Bozhena's face with the back of his hand.

"My Chosen, apologies for my delinquency in carrying out your orders. But if my plan does not work, there will still be time to make use of the one we have granted ambitions. If I fail, we can still activate him. And the girl given into his holding."

"What you have done is allow my lieutenants to be killed. You have shied from my command." He made a brief dismissive gesture with the appendage that was his right hand. "The asset of which you speak has already been activated by another."

Then, the Chosen's form turned and moved away abruptly. It remained silent for a moment, stewing. Then he said in a quiet, higher pitched voice:

"You think that your plan is so much better? Very well. If you fail, you will have to answer to me and to the one who I shall ordain as my new lieutenant. No matter how intelligent you think you may be, I can assure that it will not be you."

Having spat out his judgment, the Chosen's form was abruptly gone.

Sighing, Bozhena looked over at her captive. Upon the disappearance of Ziya's form, he had gone back to slowly consuming his meal. The young woman waited until he had almost finished. At that point, she moved her form to take the deer leg's bone from his hand. Without giving the boy any opportunity to wash his hands or take in water, she rebound him to the cast-iron stone braces. She needed to get away. To be alone. To think.

The Slav exited the ruined courtyard of the castle – for she knew exactly what it was – and walked onto the shallow slope that led down toward the steep drop-off of a ravine.

She loved this place. Her mother had taken her here once, while her life was still her own. On that occasion, she had met with the owners of the castle across the way. Shortly after, she had become an unwitting servant of the Dark Prince and His six other disciples.

The blue-eyed Urumi had since come to this locale when she needed to recharge, ever since being consigned into their service by her father's deal. It galled her that Ziya had found it.

She looked back at the gates of the castle remnants that she had just exited. Then, the girl indentured to the Dark Prince looked over to the still functional one across the ravine.

Niedzica Zamek. The name of that keep wafted through her mind. Once upon a time, she had known the rulers of its walls. They had told her the last of the Inca royals had escaped there for a time, about a century beforehand. Rumor had it that they had hidden their own treasure – most likely meant to be used against the Spanish occupation of their territory – somewhere in the castle. Yet, when the castle's rulers had brought her across the ravine to the ruins of the nearby Zamek w Czorsztynie, Bozhena had discovered that the treasure lay not in the still-occupied Niedzica, but beneath the earthen courtyard where her captive now sat tied. It was not just gold, she knew now, but something far more powerful.

At least Ziya did not find that, the Urumi thought, the phrase manifesting itself in a mix of Polish and Russian. Often, Bozhena had dreamed of using the treasure to lead a rebellion and fight for her homeland. Then, her own Russified, power-crazed father had forced her to end that fantasy. Instead, it was he who had consigned Bozhena to her current vocation. In that moment, she was overcome by a wave of hatred for him. And another of resentment directed towards Ziya al-Din.

At some point, Bozhena did have to concede that Ziya was right. She was her father's daughter. Although he'd thought of himself as Russian, she had never conceived of him as such while growing up. Before her Transmutation, she had always thought of him as a Pole, who had sadly forgotten his heritage. And as a Pole herself, she reasoned, it was not wrong or shameful for her to have inherited his ambition. She resented Ziya for his position of power; if she must be consigned to this life then why could she not lead, in light of what in the Slav's mind was her clearly greater competence?

Many times since her service with the Urumi had begun, the Slav had come to this ruined castle. She had entertained a fantasy that she would dig up the Inca treasure and begin a new life somewhere – anywhere – else, in some place that was away from the Shadow Warriors or the scars of her own past. But she knew, logically, that such a course of action was impossible; the blue-eyed girl had been made an Urumi. There was no going back. Even if she attempted to flee, Ziya or whoever succeeded him could call for her or find her eventually. Just as the Chosen had done now. And – free will robbed of her by the Transmutation – Bozhena would have no choice but to obey his commands. Escape was a childish fantasy.

Bozhena looked down and noticed that several small wildflowers with yellow buds bloomed near her black-booted feet. Looking about herself to ensure that no one was present, she allowed her true form to become apparent.

Sighing, she braided some locks of hair, such that they formed a line running horizontally downward from her forehead to the nape of her neck, almost in the line of a crown's bottom edge. She knelt and pinched off several of the flowers' stems. At regular intervals, she inserted them into the folds of the braids, such that only the flowers' blooms were visible.

The Urumi stood and stared out over the ravine. The castle on its far edge was still full of human life. Behind that structure, the sun set. Its oblique angle made the chasm of the ravine appear infinitely dark and wide. The blond-haired girl could see across, but she could not help being struck by the impression that the distance which separated her from her former life was too profound to ford.

Bozhena stood that way until long after the sunset. The lights in Niedzica became visible. Then they eventually winked out. It was the middle of the night. But her appointment with the Fragment's possessor loomed.

The time of confrontation was fast approaching. One where she would be forced to consign another to the fate she now suffered. The

Slavic Urumi sighed and turned back to the ruined castle where she had left her captive.

Like it or not, Bozhena had no choice.

Six

From the carriage window, Nell could tell that she was still being driven through a large city. It was a metropolis like none she had ever seen. The streets narrowed. The edifices were ornate, many of them made of brick or marble. As the conveyance moved across the thoroughfares, the buildings slowly lowered in height and became less embellished.

This place reminded the girl of the neighborhood that she had passed through just before having to say goodbye to Stas in Bombay. Sure enough, her suspicions were confirmed when, through the window, the banks of a wide river became apparent. More buildings crowded along its farther shore.

Nell gathered her ragged resolve, remembering what had happened since she had tried to escape.

After being assailed by the dark form, Nell had come to in the same carriage where she now sat. The man she had seen next to the ghostly figure, the one who had abruptly ended her escape attempt, had been sitting diagonally across from her. He was dressed in the uniform of a high-ranking military officer.

Upon regaining her bearings, Stas's friend had asked immediately where she was being taken. The man had only glared at her in response, as if positively disgusted by her presence. The journey had continued like that for hours. Eventually, they had entered the same city where, as far as Nell could tell, she remained.

The carriage had stopped in front of a brick building. There, she had remained locked in the carriage's cabin for a few moments. Her captor had exited; he returned a few moments later, opened the door, grabbed her roughly by the wrist, and pulled her out. Nell had yelped in a mixture of surprise and pain as she tumbled to street level. She was led roughly into a reception area. At least four or five people had been present.

"Help!" she had yelled immediately. "I've been kidnapped! Again!" In response, she received only confused looks from those around her. Her captor had merely affected a placating smile.

"Not to worry. This is my friend's daughter. He got called away on assignment. I am watching her for the time being. There is no cause for alarm. She's merely not quite in her senses," her captor had told them. All present in the room either went succinctly back to their business or stared at Nell with an odd kind of pity. It made the girl stare at the ground as she was led off like a shamed prisoner.

After, the brown-haired older man continued to drag the girl upstairs. He deposited her in a bedroom and shut the door quickly behind. Nell ran towards it, in an attempt to force it open. But, just as she reached the threshold, she heard the sound of a deadbolt being locked from the outside.

Nell paced around the room, trapped. Physically, she was in decent shape. Mentally, however, she had somehow begun to question whether the current situation was her own fault. Logically, of course, she knew that it wasn't. Yet, she acutely feared that – somehow – it was.

Could she have said something else to her captor? Something that would make the man release her? Could she have tried harder to escape on the ride to her current locale? *No*, she reasoned, *that officer was watching me the entire time*. Yet the question continued to pound in her head, over and over, until it joined a chorus of criticisms:

Just now, could I have said something different in the parlor? What if it had been the wrong thing? Is this continued captivity my fault?

The thought bored into Nell's consciousness. Again, an opposing voice inside the girl's head screamed that the self-accusation had no factual basis. As far as Nell had been concerned, screaming that she had been kidnapped again was the logical thing to do. This was the third time in as many years that Nell had found herself in such a situation. It was beginning to seem almost normal. It had even been fun on some occasions, in a weird way. But not this time. This time had been the worst of all, by far.

In Africa, she and Stas had been together. Despite her initial fear, they had escaped their captors. Nell had largely fond memories of the experience, not including a few spells of sickness. She mostly recalled time spent with Kali and Mea and of riding atop the elephant King. In

India, her captivity had been terrifying. But it had also been relatively short. At least she had been around those who had taken her. Some, like Malka, even betrayed an odd sympathy. She had also been able to escape, twice. But, it was Stas who had finally rescued her.

England had been different. The girl had spent months locked in a room. Everyone refused to speak to her. She had begun to question whether there was something wrong with her. Finally, her porter had agreed to aid her. For a brief moment, she'd thought that escape was at hand.

However, her plans had been thwarted, dashed by a black figure that she could scarcely bring herself to believe that she saw. After, Nell had awoken in the carriage. And again, nobody would speak to her. It was as if the escape attempt had been a fitful dream.

As she paced in her locked room, Nell's mind entertained the thoughts that she had refused to countenance outright until that moment.

What if it's me? Maybe I am crazy!

Nell stopped her pacing. She stood in the middle of the room, which contained light, plain brass fittings and a similar brass-framed bed. The girl remained unmoving, pondering her fears for at least twenty minutes.

What if. Those words prefaced most of the thoughts that ran through her mind in those minutes:

What if my life in Africa never happened? What if the things I saw in India were figments of my imagination? What if I've never even been there and this is all just in my head? That would explain why people won't talk to me. What if I really am the crazy daughter of that man's friend and I'm too far gone to know it?

Nell looked around the room. As she did so, she began to notice – in a disconnected manner – that her body had begun to hyperventilate. She tried to search for something – anything – that she could latch onto to allow herself to believe that her old life was real.

"Stas," she whispered aloud between gulps of air. "You will find me. You promised that we would see each other again. We will." In that moment, Nell made a promise of her own. She would continue trying to escape, no matter what. Maybe, confidence was a choice.

Still breathing heavily, the dark blond-haired girl looked around her newest cage. A window lay next to the left side of the bed. Under it, a lamp rested on a bedside table. Nell moved towards the aperture. She tried to open it, but it was locked. Immediately she grabbed the lamp and thrust it, shade first, into the windowpane. The glass shattered completely. The girl had been about to climb out when the door to the room burst open. Two porters grabbed her.

After that, Nell had spent at least the next week tied to the bed frame. She had been unbound twice per day, by the same military-dressed man who had apparently kidnapped her. She was allowed to eat and relieve herself. The man never spoke to her or responded to her questions, before retying her and leaving the room.

Some days were better than others. There were times Nell feared that she might truly be mentally ill, that her past was an illusion, that she wasn't who she thought she was. Stas's promise was the only thing that allowed the girl to keep her resolve.

He saved me in Africa. He saved me in India. And he will save me here.

Then, only a few moments ago, the brown-haired officer had entered the room. He untied her and led the girl back into the carriage where Nell found herself now.

As the conveyance stopped presently, Nell perceived that they were in front of a medium-sized ship. It sat docked on a quay next to one side of the river. A younger man in a much less ornate uniform opened the door.

"Your Grace. We have been expecting you," the younger, bull-nosed man said. "If you will follow me, I will have you on board in no time."

"Very well," Nell's kidnapper drawled.

Again, he grabbed the girl's wrist and pulled her onto the wooden boardwalk. She landed heavily, tripping as she did so, and was pulled toward the boat. The trio had almost arrived at the ship's port of embark when the young man asked a question.

"Just one thing, Your Grace: may I see your passports?"

"I would not worry. Everything is in order," the Lord Major rasped.

"Um, but, Sir? It's just routine procedure. After all, you are traveling outside of the United Kingdom and Her Majesty's Dominions."

"Are you questioning me, *Private*?"

"Um … no, Sir. Of course not, Sir."

The man flinched, visibly nervous. After a beat, he continued.

"But you will need to show your tickets."

"That," the knight said, "will be no problem." Nell watched as her captor removed two pieces of paper from his inner coat pocket. He handed them to the other military man. In response, the younger of the two men pointed to a pair of civilian porters. They were situated just inside of the ship's embarkation area.

"I'm sorry, Your Grace, but you'll have to show your proof of passage to them; I am here to see you off. But, you remember, this is a British-run service. However, it is private, Sir. I can only be present as a courtesy."

The older man glared at him for a while. Eventually, he replied.

"Oh, very well." Still holding Nell by the wrist, he moved towards the two porters and handed the one on the left the tickets with his free hand.

The porter inspected the travel documents. Then the man confirmed:

"So, you both will be traveling on with us to Danzig?"

"That is correct."

Nell began to squirm. The name of the place was vaguely familiar to her from her study of the atlas during her captivity. Whether she could place that name or not, she knew that she did not want to go there with this man.

The porter took in their image. A man in full military regalia, paired with maybe a ten-year-old girl who clearly did not want to be there.

"May I see your passports, please?"

In response, Nell struggled even more violently.

"I…," the aristocrat bellowed, pausing momentarily as he attempted to rein in his charge. Then, having gotten a more secure hold on Nell, he continued: "I am the Lord John Richard Pluckett of

Yorkshire, Knight of Her Majesty the Queen and Major in Her Royal Army. Do you question my integrity?"

The porter seemed cowed.

"No, Sir, of course not, Sir. If you will follow me, I will show you to your…."

"Wait!" Nell screamed. "Everything is not all right. This man has kidnapped me. I don't know where he's taking me. Please, help me!"

The porters stared at her for a fraction of a moment before the Lord Major could interject. Again, he told the same story.

"She is the daughter of a friend. Not right in the head. I promised that I would look after her while he went on tour to India. I didn't expect to have to go abroad either, during that time. She does not have a passport. I will handle things when we get to Danzig."

"No!" Nell began to scream.

The two porters looked from the girl to the man with the much more prestigious position. Eventually, the one who had taken the tickets made a decision.

"Very well, Your Grace. Follow me."

Pluckett did so, still dragging Nell by the wrist. Her head remained bowed towards the ground. As they walked along the ship's corridor, Nell whispered to herself: "Stas will come for me. He will. He promised."

She did so, over and over again, until the porter stopped in front of a doorway. Inside was a well-appointed stateroom. The shipping line employee motioned for Nell to move inside. Seeing no alternative, the girl did so. Yet, as she was about to pass the threshold, her captor grabbed her by the shoulder.

The brown-haired man knelt down, so that he was facing his captive directly in the eye.

"Whoever it is you think will find you, I can assure you: he will not."

Nell stared directly back at him. Desperately attempting to display none of the terror she felt.

"He will," she whispered shakily.

"He will not."

"Why?" Nell's voice quaked with fear.

"Because," the man whispered in a voice only loud enough for the two of them to hear, "everyone but me thinks you are dead." His voice carried a timbre of victory.

Nell took in a sharp breath as the uniformed man shut the door, with her standing just inside the threshold. After a second, she heard the sound of a key locking the room from the outside. She looked around her quarters desperately.

Could it be that what her captor had said was true? The thought was too awful for Nell to contemplate, even as she did so. She was stuck here; now, she had confirmation. Stas was not on the way. There was no end in sight.

Standing inches in front of a locked door, Nell was headed for a mysterious location, the name of which she could not place. She began to cry. Sobs wracked her body until she collapsed onto the floor where she had been standing. This time, the unconsciousness of sleep took her into its calming embrace.

Seven

The streets were crowded. As she threaded her way down packed avenues, Malka found herself a bit perturbed. She was not used to crowded cities. Back in India, the leader of the Thags had been slightly overwhelmed by the centers of Madras and Calcutta.

New York City, on the other hand, was like nothing she had ever seen. It was packed with people. They were forceful, unyielding. It was as if they all had somewhere to be, but they were not looking forward to getting there. Individuals of every shape, color, and size surrounded the girl. Many of them spoke languages other than English, which she had thought before coming here was the native language of this region's people.

Unused to the crowdedness, Malka moved forward, steeling her resolve. She was dressed in a flowing black dress, made of slightly reflective material. It billowed outward as it flowed down her body. Her hands sported white gloves, the fabric of which extended almost up to her elbows.

Her sash lay draped in a diagonal manner across her torso, running from her left shoulder to her opposite hip. The subcontinental native felt less than comfortable in this garb. However, it was what Liza had strongly advised her to purchase upon her arrival. The felinoid told her it would help her blend in. That way, she could avoid questioning. Despite her skin tone, the trappings helped her appear as a member of the upper strata of society in this locale.

Still, some had looked at her. They had stared at Malka, as if questioning what the Thag was doing in those clothes. Yet most passersby had directed their attention toward the black-haired, milk-skinned girl who walked closely behind her. Despite her advice to the Thag, Liza had remained in her human form, clothed as she always was: in her tight-fitting, two-piece black jumpsuit. Some bystanders leered at her. Others stared as if scandalized.

Liza glared right back at them.

Malka was unsure of the reason for Liza's show of nonconformity. On the trains that had brought them here, she had remained almost exclusively in her quadruped form. The Thag had excelled in her work in the engine compartment. Both conductors she'd worked for had offered her a permanent position, even a promotion, should the girl decide to stay. Yet, for the blue-eyed girl, it had only been a means to an end. She had declined. The nephew of the railroad tycoon had asked her to remain as well. He'd returned her saddlebags, shortly after the train pulled into Grand Central Station. Still, taking the belongings that were now hers, the Thag demurred.

Shortly after, Malka was able to use only a small fraction of those bags' contents to secure a suite of rooms in what she was told was the nicest hotel in this unfamiliar place. She had also spent a comparative pittance on an entire wardrobe of clothes like the ensemble she now wore. Of course, Liza had advised that she wear her sash more discreetly. But Malka refused to hide her foremost weapon in anything but plain sight. In spite of her dress, she was a Thag. It signified the hidden honor of her Sect.

Now, the disciple of Husain approached her destination. It was the place that the Urumi's note had bid her come only a few minutes from the present time. The street that Malka was on grew even busier. The girl crossed an intersection and approached the yellow facade of a large, imposing building. She stopped in front of its main entrance.

"This is it," the Thag said simply.

The blue-eyed girl and her protector regarded what they saw. It had a comparatively small doorway for the size of the building, Liza judged. The fading daylight made it appear more like a factory than the venue that the felinoid suspected it to be. She followed Malka's gaze to the glass-encased flyer that was posted next to the door. Liza knew her suspicions about where they were had become confirmed.

For her own part, the Thag took in what was written on the ornate piece of paper in that wall-mounted display case:

The Metropolitan Opera presents:
Les Pêcheurs de Perles
A work in three acts by:
G. Bizet
Starring:
Enrico Caruso As *Nadir*
Frieda Hempel As *Leïla*
Giuseppe De Luca As *Zurga*
Special performance. By invitation only.

"Hmmm. That's ... interesting?" Liza purred at length, although an edge could be detected in her voice.

"What is?" Malka started. "I was told to be here at this time. How does it matter what takes place here concurrently?"

"I don't really have a clue, Malka," the felinoid conceded. "However, we should...."

The Thag interrupted.

"'We'? I was told to come alone. That is what I'm going to do."

At this, the black-haired girl raised her countenance skyward. She barked out a sharp laugh.

"Malka. You're about to go into a showdown with at least...," she paused for a moment, looking for the proper word choice, "at least one, and probably more warriors with superhuman abilities. With what you carry, you're telling me that you don't want backup? It's nuts!"

"Maybe. But if I'm to get Henry back, it's what I must do." The Thag turned from the doorway to look Liza directly in the eye. "Wait for me here."

Again, the felinoid appeared positively annoyed.

"If you really expect me to stand in the middle of a busy sidewalk full of pushy people for the next few hours, then you've got another thing coming."

"Why not? I'll go in. Take Henry from them and be gone. Who said anything about hours?"

Liza shook her head in an exasperated manner.

"For one thing. That does." She pointed to the playbill that hung in the glass box on the wall. "And, for another? Do you honestly think that they're going to just let you walk out with the kid scot-free?"

"No. They will not; I cannot give up what I carry. I will likely have to fight them. But, your coming with me might antagonize them further."

"Damn it, Malka." Liza grated, almost below her breath as the busy street traffic bustled behind her. Finally, the black-haired girl threw up her hands. Then, she returned to a louder register.

"Fine. Whatever. Do it your way." The Society's protector stalked away, shaking her head as if in disbelief. She turned left, around the corner of the building that the Thag was about to enter.

Turning, Malka pushed on one of the main doors. It gave way, inwards, creaking as it did so.

With that, the would-be rescuer of a lost soul stepped into the dwelling's comparative darkness.

***

Malka's surroundings took form soon after she had entered the building. The entrance hall was surprisingly small, given the size of the edifice. The Thag looked around. She wondered what would be expected of her next. Beyond a command to meet in this place, she possessed no further information. The dark figure who'd taken Henry had given no instructions, other than orders to come alone.

Noticing a split stairwell, Husain's protégé made a decision. The blue-eyed girl started toward the left of the two staircases. She did not know what was to be found beyond. Malka had just about reached the stairs' banister when she heard a voice call after her:

"And just where do you think you're going?"

The sentence belonged to a woman sitting behind a wood and glass partition embedded into the wall on the left side of the entry chamber. Older, her face was beginning to show signs of wrinkles. The woman's hair was a very light shade of blond, as if in transition from its former luster to a dull gray. Still, the words were spoken with authority.

Malka turned. The woman's mien held a certain familiarity, but the blue-eyed girl couldn't quite place it.

"Is there a problem?" the follower of Shakti asked.

"No. There's no problem. Unless, by problem you mean your barging in here and trying to just walk up the steps to the boxes."

Malka moved toward the indented enclosure where the woman sat. The girl did not do so with malice. Instead, her mien was calm.

"I require access to the event taking place tonight," the blue-eyed girl stated. The sentence was not loaded with the charge of a demand. It was a simple statement.

The older woman on the other side of the glass partition pursed her lips.

"Did you *not* see the playbill outside? Today there isn't any opera."

With that, the gray-haired woman directed her head downward. She appeared to examine papers while treating the Thag's presence as if it were no longer extant. Eventually, Malka removed a piece of bloodstained parchment from her satchel, which lay on the same side of her torso as the sash's bottom end. The girl pushed it through the small opening in the glass partition that separated her from the prying lady.

"The flyer outside states that entrance is granted by invite. I have been given this." They were not counter-accusations. Malka meant her statements in confusion. The camp-raised youth was not used to receiving this level of suspicion from those who did not already have knowledge of her background.

The woman snatched the paper. It was as if she had been annoyed by Malka's action. Then, upon taking in the parchment's content, her eyes widened, as if making a show of recognition. She looked up. An odd smile crossed her features.

"You're Malka."

"Yes."

"Come, this way." The woman stood from her seated position. She opened a door that stood close by in the booth's wall, behind her. A few seconds later, she exited from what appeared to be a concealed passageway in the entrance hall's left wall. Oddly, Malka noted that the older lady wore pants, like the Thag herself preferred to do.

The woman approached the camp native in an unthreatening manner. As she approached, the older woman pressed a new piece of paper into the hands of Husain's protégée.

"You've got the best seat in the house. It's acoustically *perfect*."

Malka stared at the woman, as if deciding what to make of her manically enthusiastic smile.

"Very well."

The woman pointed to a wide opening in between the split staircases, one of which Malka had just tried to climb.

"Just start by going through there."

"Thank you," Malka stated.

Inclining her head in a curt gesture of acknowledgement, the once blond-haired woman turned and headed back toward the section of wall that would allow her to access the booth.

For her part, Malka moved toward the darker opening between the stairs. It was flanked by luxurious fabrics. Yet, she could not see beyond its aperture.

The subcontinental girl passed the threshold. She perceived a wall directly in front of her and turned left. It was the only direction available to her. The Thag's surroundings were characterized by dark wood-paneled walls, deep red carpeting, and dim gas lighting. Again she turned, and turned. Then she turned again. Right. Left. Then another left. It was as if she had entered a maze. Despite her foreign surroundings, there was some familiarity about the pattern that clawed at the back of the Thag's mind.

Malka kept going. The blue-eyed one kept hoping that she would notice some pattern or any alternative route. But there was none. Then, she arrived at a closed double doorway. She pulled, pushed on the handles. They refused to budge. The Thag tried multiple times. It was to no avail.

Then, it hit her.

The route. The series of turns she had taken was exactly the same as the one her Master had shown her on the night he had taken his protégée to witness the magic of the Invisible Circus, and later, when she and Stas had gained entrance to it from Madras.

Looking around, Malka perceived what she was looking for; a red tassel, its fabric streaked with gold, lay to the left side of the closed threshold. Malka reached out with her nearest hand. She pulled it. The doors opened inward.

The Thag entered a large multi-story theater. On the far end was a stage. Its curtains were up. Beneath the front end of it, an orchestra could be heard tuning itself to the concert B-flat of an oboe. She checked the piece of parchment that the peculiar, older lady had given her. From a quick perusal of the letters affixed to the seat sides, the Thag determined that the one assigned her was near the center of the auditorium. She decided immediately not to risk sitting in that space. The sinking feeling in her stomach told her: those who had invited her might just expect her to do so. She knew that they were watching her. Felt it in her stomach.

Besides, seeing that she appeared to be the only one in the auditorium, the Thag did not expect to have to quarrel over choosing her place of waiting for her assailants' arrival. Electing not to move beyond the edge of the theater's balconies, the girl chose a seat on the left of the aisle.

The music began soon as Malka ensconced herself.

It was calm, slow. Almost soothing. Yet, the Thag found herself looking around nervously, as if for something to do. Or, to address possible threats that may become apparent. For her, the two things had become one and the same.

Then, something fell into her lap. Immediately, it jumped to the ground and then into the seat next to the camp-raised girl. Its form grew. In the next fraction of a second, Liza – in her humanoid form – sat next to Shakti's disciple.

"I told you not to come," Malka grated.

"Right. Does that mean you expected me to listen?"

"How did you get in here? I needed to placate some woman at the entrance simply to gain access. I was invited."

"I have my own ways of entering the Invisible Circus, thank you very much. I'll take it you did at least notice I entered from above?"

"Yes. I did," Malka whispered in annoyance rather than acknowledgement. "But they told me to come alone. Your presence here complicates matters."

Liza harrumphed.

"Please. They don't even know I'm here, Malka. Besides, you know I have a certain way of making myself discreet. I've been assigned to protect you. And, that's damn well what I'm going to do."

"Fine," the Thag fumed. "Then why make yourself apparent?"

"Don't look at me. It's you who decided to hide under the first loge."

The felinoid turned to cast a sidelong glance at Malka.

"Don't worry, I think they don't even know I'm here, yet."

Malka craned her neck to regard the one sitting next to her. After a moment she replied, although still not as if particularly enthused: "Very well."

Then the Thag turned her attention to the stage. The music continued to undulate in a fluid manner. Onstage, performers had begun to enter. It became apparent that they were supposed to be the inhabitants of a small village.

"What is this anyway?"

Liza exhaled sharply. It was an indication of ironic jocularity.

"This," the felinoid whispered, "is opera." There was a rare hint of appreciation in her voice. The Thag turned to regard the milk-skinned girl once again. Malka's features betrayed consternation rather than confusion.

"We are simply expected to sit here?"

"Yep. That's pretty much the idea."

"For how long?"

An exasperated sigh.

"As I tried telling you outside just now, about three hours."

"I am supposed to sit in a darkened room? Waiting for them to attack for the next three hours?"

Liza directed her gaze toward the stage.

"Yep. Pretty much. Might as well get some culture while waiting for the carnage, though," the felinoid whispered.

"I would truly feel better if I had an enemy to fight." The Thag blurted the statement as if surprised by it herself.

"Just be grateful, Malka," the green-eyed girl responded beneath her breath. "If the Urumi really wanted to be sadists, they could have made you sit through the entire *Ring of the Nibelung* before making a move. Now, it's time I made myself discreet."

In an instant, a black cat sat where the bipedal form of Liza had been. It jumped down and hid itself under a row of seating a few paces hence.

Onstage, a man walked among the villagers. He was clearly meant to be their leader. They opened their mouths in melodic praise.

With that, the work began.

Eight

At first, Malka was annoyed. She could not understand the language in which the players sung. Though, she could glean clearly that the plot was supposed to be set in South Asia. More specifically, on an island near where she had been raised.

The scenery and clothing were inaccurate. They looked nothing like the place where she had grown up. Nor did it resemble the places, such as Madras or Calcutta, which she had visited since being allowed to leave the Thags' camp. The action unfolded. Malka could detect mentions made of the gods that she worshiped, but nothing resembled any religious ceremony that the Thag had witnessed.

They have conquered us. Imposed their way of living and thinking. This is the image they have of us? Husain's disciple thought with disgust. *That we are simpletons? Good for nothing but a bit of diversion?*

The blue-eyed girl continued to watch the work. The thought came into her head that many in her own camp had never accepted her either. She pushed it away. Sitting alone in an uncrowded theater, it at least made her feel marginally better to be part of any group. At the same time, she knew that the claim was a falsehood. Beyond the pronouncements of Husain, her standing as a Thag had been specious at best in the eyes of the others. Still, the closest thing Malka ever had to a people was dead because of her revelations. She now found herself in a foreign land, removed from the place where she had been held captive for years.

Malka observed the plot develop. The lyrics mentioned the names of the gods she knew – including that of the one who served as husband to the Goddess she served. It all seemed so pointless and insulting: a camp of villagers, engaged in a somehow dangerous task, portrayed in such a sensationalized manner. The dark tan-skinned girl could not push away the feeling that it mocked – demeaned – a way of life that she had known.

Then, it happened.

The melody of the strings became prime for a moment. They moved in groupings of four notes, upwards, before resetting and beginning their ascent again. A man with a higher voice – whom Malka now surmised to be one of the main characters – joined in. His voice rose, fell briefly, then it crested even higher.

Even with the cacophony taking place in front her, the Thag's highly trained ears discerned the sound of air moving. Malka stood, whirling, in one motion. She faced what confronted her from behind. It was one of the Shadow Warriors. The Urumi perched its dark form on top of one of the seat backs towards the rear of the theater. Malka could also see Henry. He was behind the figure, fastened to one of the seat rows by a length of what looked to be well-used rope.

Being at full alert, Malka took in all of this after only a fraction of a second. She jammed the balls of her feet into the ground and leapt into midair. She angled her center of gravity backward just a bit. It was exactly enough so that she landed, balanced, on the seat backs of the row nearest her. The half-breed's back faced toward the stage as she addressed her enemy.

"Release him." Malka noticed that her own statement seemed to coincide with the end of the higher-voiced performer's singing. The other performer began to respond. At the same time, the Urumi's own voice barely overwhelmed it.

"Give us the Fragment. Surrender yourself to our fate, to that which he who sired you has consigned. Or, consent, as he did, for the one who now stands bound behind, should you wish to free yourself from his agreement."

"No. You lie. My father is dead," the Thag responded. Again the girl noted that her own statement coincided with the continued singing of the higher-voiced performer. The one with the deeper register seemed to continue, mirroring the voice of the dark figure.

"Your father lives." The words struck Malka to the core. Could it be true? Her own father – her real father – still alive? For a second, the half-breed's head was overwhelmed with the audacity of hope. The Shadow Warrior spoke again:

"He has consigned you to our service," the figure's deep voice thundered, softly.

Noticing that the higher pitched singer's voice had begun to resonate again, the Thag marshaled her response.

"Not true," the blue-eyed girl began. Pausing, she realized that she did not know what to concretely accuse the dark form of dissembling about. She made a decision and continued, quickly. "He is dead. Even if he were alive, my father would never have done such a thing. He loved my mother. He died to be with her."

The higher voiced performer's statement continued for a moment before the dark figure responded. Its own rumbling coincided with the deeper-voiced one onstage.

"It is so. Your father lives. He left your mother, you and your sister. He used you to gain our favor."

"Assuming I believe you, which I do not," Malka's voice shook with an angry incredulity, "you approve of this?"

"I did not say that."

"And yet you stand here ready to oppose me? To make me – and the one I protect – become pawns in a world that you would create in your own twisted image in defiance of Shakti's will? My Sect exists to fight you."

"I do," the voice thundered, though there was a certain undercurrent. It suggested self-chafing, rather than affirmation of its own cause.

Onstage, the lower-voiced of the two continued its own melodic monologue. Then the higher one began again. The two dialogued back and forth.

"You," Malka spat. "You killed my sister. Why should I believe you?"

"I did no such thing. Now, comply," the figure responded.

"You. You did. It was you who delivered her into the hands of that despicable Prince Lubomirski. All my life, I told myself that she was the one chance I had to belong. I saw nothing of myself in Antonia when I met her. But, still. It *was* you, wasn't it?"

As she said the words, the more clarion-voiced of the two singers took the melodic lead.

"I did nothing of the kind," the Urumi voiced briefly.

"Then one of you. There is no other explanation."

"If so, I … I had no knowledge." Its deep voice seemed to state the contention simply. The lower-registered singer finished the response.

The higher-singing voice of the two onstage took the lead again. The Thag felt oddly buoyed by it. Her blue eyes narrowed at the dark form.

"Give me one reason why I should believe that."

The lower voice again, briefly.

"I am not able to give you any. Yield."

"No," the leader of Shakti's cult replied. As she did so, her left hand closed around the sash that lay draped over her torso in a blur of motion. Her right went to the ax that lay hidden in the folds of her dress. Simultaneously, the Urumi's form moved. A long, flexible metal blade emerged from its borders. It moved up and then downwards toward the Thag. Malka threw the ax towards the dark figure. Her action forced the Shadow Warrior to alter the blade's trajectory in order to intercept it.

In that moment, the two voices onstage seemed to culminate. They sung as one. Their tones harmonized with one another. Two pieces of metal came into contact. Sparks flew as they hit. The impact knocked the trajectory of both lethal implements awry. The ax buried itself deep into a panel wall. The blade of the Urumi continued downward. It slashed through the tops of at least four seat backs before it moved back up into the air. It became redirected at Malka.

Onstage, the voices continued to harmonize. Affixing the brass doorknob to the far end of her sash, the Thag prepared for battle. She threw the weighted length of fabric towards the blade that assailed her existence. Time and again it would wrap around the metal weapon. Time and again the dark form would pull it towards the Thag. Then, upwards, both slackening the fabric of the sash while freeing the blade at the same time. They seemed to move almost in a rhythm to the music. Jumping back and forth, from seat back to seat back as they did so. Throw. Leap forward. Parry. Pullback. Jump forward. Slash.

Always, the two weapons met as the two singers onstage reached their high notes. Malka continued to fight this way with the dark figure for at least – it seemed to her – a few moments. As they did so, the battleground became more uneven. The Urumi's blade seemed to mutilate at least a few seats each time they repeated the pattern of feints.

A break in the music startled them both.

It had come just as the weighted sash wrapped itself around the Urumi's blade. The Shadow Warrior had again slackened Malka's sash by throwing the blade out towards her. The rupture in the music made the figure stop. Taking advantage of the sudden momentum, the Thag pulled. The Urumi's form seemed to teeter on the seat back on which it stood. Noting this, the blue-eyed girl withdrew the jewel-encrusted dagger from the sleeve of one of her boots, which she still wore beneath the black dress.

She threw. The Thag was aware that the two performers no longer sung in unison.

The Urumi leapt out of the way. It launched itself backward over at least three rows of seats. The figure landed with a forceful jerk on its blade. Her assailant's weapon still connected with the sash, the gesture threw Malka off balance. If it had not been for her extensive training with Husain, the girl would have fallen completely off the seat backs.

The lower voiced man's melody took the ascendant again. The dark form recognized opportunity. Its quarry appeared momentarily disoriented. The figure disentangled its weapon from the bloodred sash's fetters.

The two voices began to dialogue more quickly. The Shadow Warrior brought the blade down towards the one who carried the Fragment. Malka was still attempting to regain her balance. There was nothing she could do. Husain's protégée prepared for the searing pain of metal to flay her.

It never did.

When Malka regained her balance, she looked up to see a blond girl where the dark figure had been balanced. She was about the same age as Malka. But, the girl wore almost priestly black robes. Her face displayed scars in multiple places. Her nose was crooked.

The figure that revealed itself to the Thag appeared deformed beyond her years. Yet, the braided hair, interspersed at regular intervals with small yellow flowers, drew an odd contrast with her battered appearance and severe garb.

Then, Malka noticed that the figure's face was wet. Her assailant was crying. Tears flowed silently from her intense, blue eyes.

The follower of Shakti stood ready. She was not sure what to make of the situation.

It was the blond-haired Urumi who moved first. Suddenly, she flung her blade outward as far as she could, toward the back end of the theater. During the entire battle, it had moved so fast that it would have appeared only as a blur to the untrained eye. The tip of its blade cut the rope that bound Henry to the theater seats.

The Thag watched in horror as it then closed around the brown-haired youth's arms and torso. There was nothing she could do. The range of the Urumi's blade eclipsed that of her sash. The blue-eyed girl had used the two projectile weapons that she carried. She was powerless. All Malka could do was wait for her opponent's blade to slice through Henry.

Onstage, the two voices harmonized again, back into the same soaring melody that they had joined in before. The blond girl's blade raised Henry into the air. The Thag noticed that its obtuse sides touched the boy. The weapon's razor-sharp edges cut deeply into Henry's arms. The Urumi flung Henry headlong into the row of seats directly in front of Malka. He collapsed. Unmoving.

"Take him!" the girl screamed. Malka noticed that it was not a shriek of anger or bellicosity. Instead, it was a half-sobbed cry of anguish.

"Take him," the slight, white-skinned girl repeated more softly. Onstage, the two voices had begun to sing as one. They were counterpointed by bursts from the brass section. The duet neared its culmination.

"Why are you doing this? You killed my sister." At this point, Malka's words were spoken more out of nonunderstanding than angst.

The blond-haired girl looked at the floor, shaking her head. She recoiled her flexible weapon.

"Take him," the one clothed in the garb of the Urumi strained. "We have more in common than you might think."

The singers culminated in a sustained note. Then, they followed it with two decidedly lower ones of the same pitch. The orchestra rose in crescendo.

"We … we will meet again," the blond-haired girl rasped. It was not a threat, but a promise made in desperation. It was more sobbed with anguish than imparted with the confidence it had been intended to carry. Then, the figure of the young girl – aged beyond her years – disappeared from Malka's glimpse.

The opera continued as she jumped down from the seating and knelt beside Henry. Grabbing him, the Thag turned his face towards hers.

"Henry? Henry! Are you all right?" Malka asked the question with a certain desperation of her own. She had taken responsibility for the youth. She did not want to think that she had failed by losing him.

"Henry! Henry!" Malka continued yelling. She barely registered that the performers onstage seemed not to take note of any of these disturbances; a woman arrived in their fictional camp.

"Malka?" the blue-eyed boy groaned after she turned him over. Blood leaked from deep cuts that slashed the flesh of both of his upper arms.

"Yes, Henry. It is me, Malka." The Thag said the words as much for herself as for her self-proclaimed charge. Simultaneously, she ripped lengths of fabric from the hems of her dress. The girl used them to bind and put pressure on the blue-eyed boy's wounds. Then she grabbed his left hand, helping the listless youth to his feet.

"We must go," she said simply, heading for the exit. Henry did not budge. Dazed, he did not seem particularly sure on his feet either.

"Um, uh, Malka. Thanks for saving me, and all that." The camp-born youth seemed to be struggling to string the words together. "Can't we stay? I've always wanted to see, um, one of…." Henry looked toward the stage, where a man seemed to profess his love to the girl, newly brought to the village. He again seemed to search for the proper words. "Uh, you know … these things."

In response, Malka pulled on his hand.

"Henry. You're bleeding."

"And from the sounds of things, you probably have an acute concussion." Liza was suddenly present in her human form. Beside them, she was seated in one of the chairs. Then the felinoid looked at the boy conspiratorially. She continued:

"Hell with it. Let's see the rest of this thing."

Henry sat down.

Liza turned to her own, still-standing charge.

"You're outvoted, Malka," the back-haired girl purred softly.

"Very well." The Thag ensconced herself heavily in the half-sliced-up seat next to Henry. "I'm really just supposed to sit here for only Shakti can tell how long?"

"Don't worry," Henry replied. "I'll, uh, wake you up when it's over if you fall asleep."

"Actually, I'm rather sure you're the one that has to worry about not doing that," Liza retorted. "Now shut up. Both of you. I want to listen."

As it turned out, none of them dozed off. Malka, for her part, found herself eerily captivated by the apparent story, if not the music. It seemed just a bit too close to home, despite the cultural inaccuracies: a small camp, devoted to its Hindu gods, involved in risky business. Struck by calamity, its end had come abruptly. It had hit Malka with the force of a thousand storms; the village leader, the one with the lower voice in the duet, destroyed his entire tribe just to save his friend and that man's charge.

After, Liza leapt to her feet, clapping thunderously. Henry, at first, had appeared unsure of what was going on, not least because there were only three of them in the entire theater. He eventually followed the felinoid's lead. He'd struggled to his feet, clapping in a limited manner with his bloodstained limbs. It was as much as his injuries would permit.

Malka sat with her mouth open. Eventually, she rested her head in both of her hands.

Softly, she wept.

Nine

The three entered their suite of rooms. It was sumptuously appointed. The floor was made of black marble, covered in places with thick oriental carpeting. Large chandeliers provided illumination. The furniture of the living room was plush, intricately carved. On either side of the parlor, a doorway led to a bedroom, each with a large four-poster canopy bed and its own full bath with sunken roman tub. The room's far end contained a wall with large windows, adorned by deep green draping.

It was after midnight. Not much could be seen through the windows. Despite the late hour, none of the three seemed to be in any condition to sleep.

After the opera had ended, Malka dried her eyes quickly. She did not want to let her protector or charge see how truly distraught the performance's content had made her. Suddenly, the stage went dark and the three turned to leave. There, they had split up. Liza mentioned something about having to use the same path to exit as well as enter the Invisible Circus. With that, she transformed into her feline form and leapt for the first balcony.

Malka, with Henry in tow, moved toward her own point of egress. Stopping briefly, she retrieved her ax and knife. Internally, she had been too fraught to do so in anything other than a mechanical way.

The girl's mind whirred with questions, the possible answers to which scared and confused her. Was her biological father alive? Had he consigned her into the eventual service of the very Shadow Warriors that it was her purpose to fight? From which she was to protect the diamond? She had been quite young when brought to the camp of the Thags. Could it be that her memories deceived her?

Then, there was the question of Malka's sister. She had always assumed that the Urumi who tracked her had somehow found that Antonia was in possession of the real diamond and delivered her into the clutches of the scheming Russian prince, Lubomirski. She would have dismissed the dark figure's statements to the contrary out of hand. But, there was its – her – odd behavior to consider. The dark warrioress had been in tears. Could it be that her statements were genuine? In a larger sense, Malka wondered if either topic truly mattered. She had been raised as a Thag. She did want revenge for her sister, but that could wait.

The prospect that terrified the Thag most of all had the least factual basis. On more than one occasion, Husain had implied that he'd known something more. Something that Malka was not yet ready to know. In others, he had been interrupted just before seeming to impart a new truth. Malka knew she had no evidence to support her worries. But, she could not get the thought out of her head when she mentally pored over the events of the opera's plot. Could it be that Husain had intentionally destroyed the Thag's camp because of her?

Just the thought shamed her to the core as she moved down the streets of New York. The walk back to the hotel took only a few moments on much less crowded midnight streets. The hotel came into view. Its name meant nothing to Malka, but she had seen Henry's eyes widen upon taking in its name: Fifth Avenue Hotel. Not bothering to question this, the Thag continued into the lobby. Liza was already waiting for them.

The entrance staff took note of their arrival with suspicion. It was not every day that one saw a dark-skinned girl, clad in a ripped and disheveled designer dress, and a filthy, blood-smeared youth, dressed in rags, enter an establishment like this during the middle of the night. However, a few words from Liza to the effect that she'd been expecting them, the production of the keys to her suite by Malka, and a high denomination banknote pressed into the palm of one of the porters, allowed them to pass with little further questioning.

As the door to the suite's living room shut, Malka walked straight to a red silk sitting chair and collapsed into it, her back hunched over. The blue-eyed girl stared at the floor. By contrast, the other two started their mouths immediately.

"That was incredible! I never thought I'd get to see an opera myself, growing up like I did. But, um, Malka? How can you move so fast? I mean, um, none of what happened really makes that much sense when you think about it. How are they going to clean up all the damage to a major theater before morning without someone noticing?

And how come the singers didn't notice what was going on?" Getting no response from Malka, the blue-eyed boy turned to the felinoid. "And what was that you said about having to get out...."

"Henry? Stop. You do get that we weren't really in the Met, right? The Invisible Circus. It's like this other dimensional space that the Urumi control. Weird stuff happens there and none of that is my problem or even the point."

Liza turned to the darker skinned girl.

"How are you just sitting there, Malka? After what we've heard, we have a potentially fatal strategic weakness on our hands and...."

"Um, Liza? But, how do you know it's not the point?" Henry was not done. "That thing held me captive for months. Then it just releases me? In the middle of an entire opera, put on just for three people? I mean, guys, how does that make any...."

"Henry. Go clean up." Liza barked.

"Um, Liza. I need some answers, here. After being held captive outdoors for months, I think I...."

"Stink," the felinoid finished for him. "And you have blood all over yourself. Go take a shower. Now."

"Fine. Great to be around you again." Rolling his eyes, the boy turned and moved toward one of the bedroom doors. "Go through all this trouble to rescue me. And you're still just ordering me around."

"Truth be told, I argued against it," the black-haired girl said after him.

"Um, thanks, Liza. I feel so much better knowing that," Henry's voice called back from the next room. With Malka's human charge momentarily out of the picture, Liza turned herself to the evidently sulking Thag.

"All right, Malka. What the hell's wrong now? You got your little friend back. And you got incredibly lucky. Again." Sighing in frustration, the felinoid rolled her eyes. "Whatever it is, I don't care. You need to put a lid on it and focus on coming up with a new plan of action that takes into account what we've just heard."

Without looking up, Malka responded.

"Henry does not know how the Invisible Circus functions. Neither, completely, do we."

"Malka, I am *not* talking about what Henry just said, and you damn well know it. If your father is alive, and if he did what that thing said he did, we have a *big* problem."

"We do not know for certain that either is true. Besides, I cannot see how it alters the options available to us."

"*What?*" the green-eyed girl was incredulous. "Malka, I cannot believe that even you would think this way. This is huge! Even if you manage to hide the Fragment somewhere, if they can make you undergo their Transmutation – make you one of them – then, they will be able to simply order you to hand it over, or lead them to it. It's time to put aside whatever internal baggage is weighing down your head and focus on a major tactical reassessment."

Ignoring the felinoid's insult, Malka continued to sit, as if lost in thought. Seeing this, Liza continued:

"And then there's the one intelligent point Henry just brought up. Why did that thing just release him? And, more to the point, why and how did it show itself?"

This time Malka spoke.

"She seemed to be in distress. She did say that we appear to have that in common, at least."

"It was having a breakdown? Really? Wow, Malka. That's a great interpretation of the situation. I'd say it's dead on. Except for one tiny thing: Urumi have no free will! There is no way it can act on having a mental breakdown. They must be playing some type of deep game. Maybe they sent Henry back to steal the Fragment for them, in exchange for his freedom."

"He would never do that," Malka snapped.

"How the hell are you so certain!" the felinoid exploded. "He already tried to escape twice."

"We told him who the Urumi are. He knows the consequences, Liza."

The felinoid shook her head as if unable to believe what she was hearing.

"I know that, but at least I'm considering all the options. You're really just going to choose to believe whatever that thing told you and accept its goodwill?"

Malka looked up, fixing her gaze directly on her protector.

"I never said that."

"Fine then, Malka. Out with it. If you're not taking its comments all personal, then what?"

"It's…," Malka hesitated. It was as if she knew how the felinoid would react to what she was about to say. Sighing, she continued. "The opera. It reminds me of my role in what happened to the Thags. They were right. I betrayed them."

At this, Liza snorted.

"That's it? You're wallowing in self-blame over what happened to your people because of *an opera plot*? Do you even hear yourself? Malka, it's an opera. Besides, the Urumi, whose motives you inexplicably seem to be taking at face value, probably chose it precisely *because* it wanted to rattle your cage."

"Logically, Liza, I am aware of that. But still, I cannot shake the feeling…."

"You're being paranoid, Malka. Again. And, as usual, it's about all the things you don't need to get paranoid over."

"Yeah. Liza's right. Half the plot didn't even, um, you know, make much sense." It was Henry, having returned from his ablutions and the tending of his wounds. He now wore a clean white button-down shirt and tan trousers.

"Uh, while you guys were fighting it out over me, I got to watch the whole thing. You know, the duet? In it, the two main characters swear to remain brothers-in-arms and vow not to pursue the same woman, who ends up being brought to the camp to appease the gods.

"Except, it turns out that the tenor is lying through his teeth. He and the woman are caught together. The baritone, who plays the village leader, orders them killed. Later on, he changes his mind, apparently because he recognizes some necklace he once bestowed upon her. The villagers decide they want the two executed anyway because some random storm hits their settlement and the camp's zealous priest encourages them to blame the woman for it. So, the leader decides that the best possible way out of this predicament is to cover up the pair's escape through destroying his entire village. The end.

"So, I mean," Henry continued, "I liked the music, but half the plot points don't even make sense. If they were trying to rattle you, Malka, I don't think showing you that opera was the best way to do it."

As Henry summarized the plot, Malka seemed to stare more and more off into space. Her eyes grew wide in realization. A few seconds of silence filled the room as the boy finished.

"Malka? What is it?" Seeing her reaction, the brown-haired boy pressed.

Without looking at either of the other two in the room, the Thag replied in a small voice. The words were more whispered to herself than spoken to the entire room:

"He knew."

"Well, that certainly speaks volumes. *Who* knew *what*?" It was Liza, apparently unsatisfied with the specificity of her charge's statements.

Malka did not respond immediately. A painful memory unfolded in her mind's eye. It was of the last conversation she'd ever had with her Master. The time she had taken that final step, which condemned the Thags to death.

***

The horse burst into the Thag's camp. It did not slow down as it did so. The two sentries on either side of the village's only entrance moved to pursue. Their efforts proved to be less than effective. The defenses of Shakti's Sect lay in stealth and discreetness, not brute responsiveness.

Husain's protégée kept the horse at a gallop. She rode the animal down the camp's most used lane. Other villagers took notice of the commotion. They moved to follow the horse as it moved to the village's rough center.

Much of the village began to surround the half-breed. Malka stopped her mount in front of the Sect leader's hut. She jumped off the horse and ran into the dwelling, not even bothering to knock.

"Master! I have returned!" the one ordained to succeed him called out.

It was only after giving her initial statement that Malka realized: Husain already sat before her. The village leader was in the sitting area of his hut, seeming to be engaged in an earnest conversation with Hamda, his wife.

"Yes, Malka," her Master replied. "It has been some time. Our quests do not always go as planned. However, I take it that you have neutralized what threatens our Sect?"

Hamda got up. The woman moved to the cooking area of the hut. At the same time, Malka stared at the dwelling's earthen floor.

"Master. Forgive me. I have faltered. Again. I have failed to eliminate my targets, the ones who were seeking this village, with suspicion of its true nature."

Husain regarded her with his usual impenetrable sereneness. His protégée took in a few sharp breaths in anticipation of her Master's response. Eventually, the man nodded slowly.

"I see. Take a seat, Malka."

The girl did so, sitting opposite her mentor.

"For what reason?" There was no hint of judgment in his voice. Instead, Husain seemed only mildly curious. Malka sat in silence for a moment, unsure of where to begin. Eventually she picked a point of departure.

"Master, you already know, at least in part. Since my appointment as your successor, my position in this camp has been controversial."

"I am aware. However, Malka, I have told you. It does not matter."

Husain's pupil shook her head sadly.

"But Master. You yourself have told me: I will never be truly one of you."

"Yes, Malka," Husain confirmed. It was far from confirmation of an accusation. His statement seemed to question why the girl thought this was a problem. The blue-eyed girl continued.

"I infiltrated their group as I was taught. I learned of their purpose and fabricated a ruse to draw them off in the opposite direction. I waited until morning, when my targets awoke, unsuspecting. It was all going according to plan. Then … it went wrong."

"What went wrong, Malka?" Again, it was a simple question. If anything, there was a trace of sad sympathy in the man's voice.

Malka paused, bowing her head at the floor in the shame of what she would have to reveal next. Especially, after Husain had granted her some measure of recognition in the camp. Confession was not her main reason for returning. Still, telling him hurt in a way that the blue-eyed girl could not quite describe.

The ordained future Master of the Thags took in a sharp breath and held it for a moment.

"I went wrong," the girl croaked in admission. "I had been about to place my sash about the neck of the first of the two. Then, he mentioned the existence of another like myself. Skin like yours, but eyes like my own. The thought that there was another out there like me. One who would not question who I was...."

The girl stopped again. She let out another shaky breath.

"It was too much to bear. So, Master, instead, I betrayed all you have taught me. I went with them. It turned out that they simply wanted to find the girl I allowed to escape our camp. I pretended to share their goal in the hope of being reunited with my sister."

Malka took in another breath of air. She prepared herself for the lie she would have to tell, if she were to have any chance of gaining the favor of her Master's aid.

"I understand that my behavior is inexcusable, traitorous. I quickly came to regret it. I accept whatever punishment you decree. But know first that during my travels with those assigned me, I discovered the existence of an object: the largest jewel known to exist. It may even be the object that Shakti seeks. In hope of redeeming myself in the eyes of Her great Sect, I pursued it with them. I decided to eliminate my targets after I'd gained it."

Lying to her Master was excruciating, even if they were mostly lies of omission. It was true that the girl had remained conflicted for most of her travels. There were times she had spent waiting, in Mungo's beachside bungalow, when she had become terrified with certainty that the Thags would come after her and that they would mete out final punishment for her latest failure. It was irrational fear. She knew that logically, yet the blue-eyed girl had felt it acutely.

Malka had learned of the existence of the diamond. Yet, she had also failed to mention that her main goal had remained finding her sister. At first, the trained killer had remained largely unconcerned with finding Stanley's friend, or the supposedly powerful object.

It was also true – at least at first – that she had merely intended to put off killing her two targets. She had rationalized that this would allow her to proceed on her escapade, while still technically obeying her Master's commands. However, the camp-raised girl had been unprepared for what had happened. The two had not accepted her, exactly. The red-haired Brit had seemed quite taken with her mystique. The other one, with the odd name, was different. He'd often regarded her with suspicion. And then with a certain contempt for what her people did, after he'd found out. Yet, she felt an odd sort of kinship with him and his situation. At home nowhere. With no people of his own. Except, apparently, for the girl Malka had saved and brought to the camp. In a sympathetic sort of way, Malka had come to envy him. At least Stanley had Nell.

Silence had fallen between the Master and his student. Husain had listened to Malka's monologue with his usual impenetrable stare. Then, he allowed a sad smile to play across his face. Malka feared that he was truly disappointed in her.

"This jewel. You do not have it." It was a statement of certainty.

"No, Master."

"Then why return? You have said yourself that you have failed in your mission. Your loyalty is already questionable. You bring back nothing to glorify Shakti's altars. Why come back now?"

The village leader's meaning sunk in. Malka knew her Master well enough by this point. He was attempting to push her towards a certain realization by calling attention to her own fears. That meant he must have a pressing point in so doing.

Malka responded. Now was the time for her most direct partial truths:

"I have returned because I require the assistance of the Sect to acquire the object. It has fallen into the hands of those whom my targets and I quested against. It is in a caravan on its way from Pondicherry to Madras. They will likely suspect what we are, if we attempt to join with their party in ambush. Our usual method will not work.

"I am here to implore you to put together a raiding party, so that we may attack directly and gain this greatest of trophies for the glorification of our Goddess. I pray that this deed will finally allow me to achieve true esteem in the eyes of my people."

Again, none of what the girl related was an outright contradiction with the truth. She did somehow hope that she could play a role in the final winning of the diamond and in the elimination of the power-mad Russian prince who sought it. She hoped that doing so would finally win her true acceptance among the camp's members. However, her other, more secret motive was to attack in the hope of securing the freedom of Nell, as well as Stanley and Mungo, with whom she had been traveling. Malka did not mention that the caravan carried a fake jewel. Or, that it had been given to her by a mysterious woman dressed in white. Malka had given the real diamond to her sister as a decoy. She'd thought no one would suspect the mild-mannered nun of complicity.

The trained thief also did not recount how, prior to returning to the camp, she had appealed to Mungo's father – Madras's police chief – to fight alongside the Thags. At least, she had secured the man's agreement that he would not arrest any of the Sect's members after the battle. But, Malka was not sure that she believed his assurances. The rather formal officer held the girl in utter contempt; after all, she had tried to kill him when they first met. But, the life of his son was at stake. That had seemed to alter his decision-making criteria.

Finally, Malka did not mention that it would be an uphill battle, though she was more than sure Husain would be aware; the tactics of the Thags were based on surprise, ambush. Any frontal assault would likely meet with a high death rate. The girl also did not mention that the party they needed to raid likely carried firearms. She had to convince her Master to agree. The stakes were too high.

Husain shook his head slightly.

"You betray us and then come back looking for an impossible redemption by begging our assistance. Interesting."

Malka sighed in resignation. It appeared that the village leader had elected not to believe her. Those she sought to rescue would be killed.

As likely would she – a punishment not undeserved. At least the diamond would remain safe.

"Very well," her Master said at length. Again, a sad smile crossed his lips. "We shall go." The Master said the words simply, hollowly, as he got up. He moved toward the hut's flap. Surprised, both by his agreement and nonjudgmental manner, Husain's protégée followed suit. Just before reaching the dwelling's point of egress, the camp leader paused, turning to Malka. His face questioned her.

"Tell me, Malka, did you ever find your sister?"

"Yes, Master. In Pondicherry." The blue-eyed girl attempted to keep her voice neutral. Did Husain suspect? Did he know that the real jewel was not in the caravan she intended to attack?

"Did you see something of her in you?" the Master asked.

"No. She was nothing like me." Abject disappointment could be heard in the half-breed's voice.

Husain nodded, knowingly.

"Why do you think that was the case?"

Malka was perplexed. Now seemed hardly the time for a philosophical discussion. However, her Master seemed determined to have it. In hindsight, it was almost as if the leader thought there would not be time later.

"Because…," the girl ventured, "our lives had been too different. Like us, she was a…," Malka paused, looking for the right word, "a priestess. For one of their gods. Yet, we could not understand each other. In the end, I feared she would not respect me if I revealed my true nature."

It was the same conundrum that she had with everyone, the half-breed reflected.

Husain nodded sagely.

"Your lives had been too different," he repeated, as if to emphasize Malka's own words. The village Master proceeded:

"Ah, Malka …." He had said those words to her many times, but now it was with a tired sigh. He looked his successor directly in the eyes.

"It is not ties of blood nor stories of history that make commonality, create trust. Shared experiences – sufferings – create belonging. If those events were not the same – not experienced together – it is unimportant. If nothing else, remember this." As the village leader spoke, his tone startled Malka. The words were formed with an intensity she was not used to hearing from the serene man.

Her Master turned resolutely. He exited into the daylight.

The first thing Malka noticed once outside was that almost half of the camp's population was already present. They surrounded the leader's hut and main temple. Zaima was standing almost directly to the right of the doorway to the leader's hut. She glowered at Malka, mouthing the Tamil word for 'traitor.' It was then that Malka realized she had been so distraught that she had elected to speak to the camp's leader in her native language, rather than English, as was their custom. Malka found herself uncertain over how much of the conversation the brown-eyed girl had managed to overhear.

Husain moved to the bell that stood outside the temple. He rang it three times. It was a sign of convocation. Quickly the crowd swelled to include the entire village. Most of those present chatted animatedly. Husain made a calming gesture with his hands and spoke in a manner that made it clear; he expected them to become silent for his announcement.

"My people!" he called. "The one ordained to become our next leader, the one upon whom I have seen fit to bestow the Sign of Aghasi, has discovered a trophy, one that will bring unparalleled glory to the Black Goddess whom we all serve. The time is now. Gaining this object will require all of our efforts. Immediately, we – every one of us – shall embark on a total quest in Shakti's name."

A commotion went up in the crowd. Malka herself was shocked. She had requested only a normal raiding party be sent with her, but what Husain called for was unprecedented. Essentially, he had just ordered the entire camp into a state of total warfare. The crowd's confusion was understandable. Some of them moved, beginning to comply. But then, a voice cut through the commotion with the force of three single words:

"I. Will. *Not!*" It was Zaima. She yelled, glaring directly at Malka and her leader. Many of the villagers remained hesitant. They turned to stare at her.

Husain briefly pursed his lips before offering a response.

"You have not been given a choice. I have called upon you to quest for Shakti. As a member of our Sect, you are bound to obey."

Most of the crowd offered up sounds of zealous approval for Husain's statement. In the past, many of them had been quick to support Zaima's agitations. Before, her arguments had been grounded in accusations of Malka's difference in origin and in common interpretations of the Thag's traditional dogma. Now, the brown-eyed girl appeared to contradict those arguments. She was refusing to quest.

Zaima remained undeterred. She moved into the empty space between Husain and his people, turning to face the assembled mass.

"I will assume that you have noticed. Our Master and his most prized pupil do not usually converse in our language. Yet, I have heard them speaking it just now." She nodded. "Yes. Our Master has called upon us to quest for our Goddess. But, on a scale never before seen in the history of our great Sect? Has he told you why?"

A rumble went up in the crowd. Shouts could be heard. Rather evenly split, some demanded more information. Others admonished that the Sect's Master owed no explanations in return for their loyalty.

Zaima continued. "He is asking all of us – as one – to follow the foremost traitor in our midst into battle!" The brown-eyed girl moved to point an arm towards Malka without turning. Her voice bore an overwhelming incredulity.

Shouting in the crowd intensified. The calls for explanation seemed to be winning out.

"It is true!" the camp native confirmed. "I heard them talking just now. The outsider disregarded her mission. Intentionally. She failed to eliminate the targets assigned to her. And now? The half-breed dares return to us? Asking our aid? I am not sure which disgusts me more: the shameless gall she has shown us in doing so, or the fact that the one meant to lead us has not only granted the request; he's responded to it by placing our entire Sect at the whim of her disposal."

Many shouts of approval could be heard at Zaima's oration.

Husain responded softly, quelling the din of the audience so that they could hear their leader's response.

"You neglect to mention, Zaima. Malka's actions led to the discovery of the object for which we now quest. If it were not for those deeds, we would not know about it. You all know, those who lead raiding parties have discretion in how their missions unfold. Malka made a command decision. Nothing more."

The assembly began to murmur, as if debating the merits of what was said.

"She failed to carry out her mission because she was thinking of leaving us! She wanted to find her true sister and betray our location!" The murmuring in the crowd grew louder.

"I never said that!" Malka screamed at her nemesis. Even in thought, she had not really considered leaving the camp permanently. Most certainly, she had never considered revealing the camp's nature. The blue-eyed girl bristled at the camp-native's accusations.

"It is true. Malka did not. Nor do I believe that she had any intention of doing so." It was Husain, offering his reassurance to the village's population. At this, most of the villagers appeared suddenly placated.

Zaima, however, was not among that number. She turned, glaring at the Master, as if attempting, one last time, to explain to him a fact so elementary that young children should possess little trouble in grasping it:

"She does not *need* to say it. She is an outsider."

"None of which negates what she has found. You do not understand what is at stake. As Master, I know what must be done, for all of our people, in Shakti's name. Come, all of you. We must all prepare." Some of the villagers began to split up. Husain began to move as well. As far as the camp was concerned, the issue was settled.

Yet, Zaima remained unmoved. She stared upwards. Her hands rested, palms exposed, at her sides. After a short, tense moment, she took voice. The village's attention returned.

"You are right, Master," the brown-eyed girl voiced a special emphasis on the title, nodding her head. "I do not understand. I never have. All of my early life, I was groomed to be your replacement. Then, she arrived. And, one day, she goes from a captive attempting escape to your protégée?" She snapped her fingers. "Just like that, my destiny was stolen from me. It was ripped away. You gave it to one whose devotion to our cause is suspect," Zaima snorted harshly. "It is not just me. Many in our camp are wary of her, but they stand ready to follow her now because you command them to do so. After all, you are 'Master.'" The camp native turned to face the crowd. She raised her voice even louder.

"I watched as you granted her accolades when she had never even left this camp. I had already quested for Shakti for many seasons by then. Oh, you may say that she has done so – once. But what has she accomplished? Your prized student brought back another outsider, who escaped and now threatens our anonymity." The brown-eyed girl turned back to Husain. "Yet you rewarded her with another mission." The brown-skinned girl continued in a lower, more mocking tone. She made a show of shrugging. "Your chosen successor failed in that, too weak to achieve her assigned targets." Zaima's voice rose again. "Is this truly who you believe is more qualified to lead our Sect than I?" She extended her hands toward Malka. "You *are* right. I do not understand."

Zaima let out a short sound that was something between a laugh and an exhalation, shaking her head, before continuing.

"What of me? I, who have worked so hard to gain your favor, who has quested for Shakti on multitudes of raids, whose loyalty to this camp remains beyond reproach? I am derided for my concerns. Overruled and overlooked." Her voice became smaller, yet it dripped with disgust. She nodded again.

"You are right, Husain. I do not understand. I do not understand because it is senseless. You have robbed me of my destiny. Given it to an outsider. No matter what she has done – what I have done – you refuse to change your mind. Now, I am expected to follow her into some great battle?"

An odd smile crossed Zaima's face as she shook her head, slowly.

"No, Husain. I tell you: I will not. If this is how you are to rule as Master, I will not follow you, either."

Zaima's voice reached a low point as she finished her tirade. The camp-native's tone seethed with hatred. Silence dominated immediately after the girl finished. Most of the assembled Thags were shocked at what, essentially, amounted to a blatant announcement of insubordination, voiced directly to the village's leader. What was more, while many in the camp remained suspicious of Malka, they had often supported Zaima out of concern for the Sect. However, in her rage, the brown-eyed girl had revealed the true source of her opposition. It rested not with concern for the Sect of Shakti and its members. It flowed from her own personal angst. None in the camp made any gesture in her support.

Finally, the Master of the Thags responded. His voice was quiet. It was also hard as cold metal.

"You will follow us, Zaima. All members of the Black Goddess's Sect will see to that. Then, it is we who shall decide what becomes of you." The Master of the Thags raised his voice to address the entire congregation.

"Enough time has been wasted. We must prepare. Come!"

The entire village leapt into action as if it were a well-oiled machine. Horses were hitched to carts. Weapons and sashes were prepared. The scale of the operation lent a certain portent to the proceedings. It was as if all of their history had led them to this day.

Older Thags and young children were loaded into the carts. The remainder of the Sect arrayed around them. No one was left behind. This was total war. Even Zaima, who found herself roped to a cart, was accounted for. Thus, the massive caravan became organized.

Malka wondered why Husain had seen it necessary to convoke the entire Sect in so grand a manner. But, too busy coordinating logistics, the blue-eyed Thag did not find the time to ask him why. She would never speak with her Master again.

With a shout from Husain, the caravan began to move; the Thags set out from their camp at high noon, bound for a meeting with destiny.

***

The mind of the Thags' last Master returned to her suite at the Fifth Avenue Hotel. Her blue eyes betrayed a certain despair. It had taken only a few seconds for the memory to flash through her consciousness.

"He knew. Husain knew they would die. Somehow, he knew what was going to happen, yet he sent them anyway. Because of me. The Thags perished because I asked him."

"That's it? That's what has you so bothered?" Liza asked. There was a trivial incredulousness in her voice.

"Liza!" Henry yelped. "I know you're ... well, that's just cruel."

Malka sat, hunched over. She looked up, sideways. Her brow was slightly furrowed. Her mouth hung open. It was as if she could not believe that even the felinoid could be so callous.

She croaked. "My Master allowed me to become responsible for my entire Sect's death and all you can say is 'That's it'?"

"Yes, that is all I have to say. I don't know why you're so worked up about this. *Of course* he knew." Liza relayed the information as if restating a widely known fact.

"I do not understand. How could he have known?" Malka sighed. She had gone back to staring at her feet.

Liza let out a quick exhalation of astonished disbelief.

"Wait.... You mean you didn't know?"

"Know what?" the Thag mumbled.

"About the Prophecy."

"Wait. What!" Henry interjected. "There's a Prophecy now?"

"One thing I've learned in this job, Henry," Liza lectured to the brown-haired youth. "When it comes to these types of mystical vendettas, life can get a bit like trash-epic fantasy fiction. There's *always* a prophecy."

"What?" Malka cast another sidelong glance up at Liza, who, once again, turned to address her charge.

"The Prophecy...?" Liza paused slightly as if this would jog her charge's memory, "of the Fragment's Restoration? No?"

Malka stared back. She shook her head slowly.

"It's been passed down from Master to Master since your cult got started?" She was still apparently trying to see if the Thag's chosen leader would remember something that she was clearly supposed to know.

Looking back down, the blue-eyed girl shook her head again.

"This Prophecy. What does it say?" It was Henry who asked the question.

Malka sat up and looked at her. Liza realized that she now had their complete attention.

"It was part of my briefing from the Society, for this lovely assignment," the felinoid began. "It was performed as one of the two main stunts, staged at the first Invisible Circus that the Thags' first Master attended. Zitar and Arunesh were there too. So we know about it."

"What was the first?" Henry intruded.

"A visual representation of the Fragment's more ancient history. Malka already told you about it. So, I know, at least, that both of you are aware of that. The second one was much less elaborate; it involved description of the future, rather than the past."

At this, the brown-haired boy opened his mouth again: "Hold on. Why's that?"

Liza hunched her shoulders and offered a three-tone grunt response.

"Anyway.... It consisted of a medium relaying her perceptions about what would happen with the Fragment. But that means all we have to rely on is that person's speech. We don't have any actual visual representations of the persons involved, nor any context about who exactly they are, or when the Prophecy's events could be expected to occur. Predictably, most other useful details were left out, too. This means there's a lot more guesswork involved. And, that the stakes are higher, because whoever interprets this Prophecy's signs correctly may manipulate circumstances to obtain the Fragment." The felinoid paused for a moment. She continued in a more exasperated tone, "And, considering the enemy we have to fight, that is also why I can't *believe* that Malka was never told!"

Liza let out a frustrated sigh. Malka continued to stare, glumly.

"Okay, so what does it say?" Henry prodded. Liza exhaled before continuing. As she talked she made quotation signs with her hands, to mark the sections she quoted from the original account.

"It holds that a future leader of 'bright eyes' will become apparent. Possessed of abilities of 'detection and speed,' this person will find the

Fragment in an 'isolated pocket.' After an ensuing struggle, the one who shall lead the 'clan of searchers' – but is not 'part of it' – will cause its 'chief hunter' to destroy that clan in order to protect what they seek from 'dark forms.' Then, travel over 'sea and land' was perceived."

"That sounds like a clear enough start," Henry offered.

"Well, yeah, but only because we already know that Malka has weirdly blue eyes. That she was brought to the camp, got picked to lead the Thags, and is sitting in a New York hotel room. Also, what's very probably the Fragment is in a bag slung over her shoulder as she wallows over the loss of her Sect. Now, if you had to notice all of these things, or just one or two of them as they became apparent, could you?"

"Good point. If I didn't know that already, it would be a lot harder."

"Eventually, after many struggles, the 'haunted keepers of the stone,' who will know each other in a place of 'common difference,' will arrive in a 'bright pinnacle of darkness.' Apparently, that's where the 'wooden object' that can either unleash or secure the Fragment's full power lies. There, a battle will ensue between the 'dark forms' and those of 'common belief and tribulation.' The outcome will result in 'destructive re-creation' and 'inclusive divisions.'" The felinoid looked pointedly at the brown-haired youth.

"See what I mean?" she asked.

"So? Who wins?"

"Who will actually prevail...," Liza blew out a long breath, flinging her hands out to her sides in a what-did-you-expect type of gesture, "is not known." The felinoid lazily allowed her hands to flop back down to her sides.

"Great," replied Henry. "So what do we have? 'Dark forms.' That pretty clearly refers to these Urumi."

Liza nodded.

"But what about some of the others: those of 'common belief and tribulation'? Who does that refer to, the Urumi as well? And, what about 'haunted keepers of the stone'? That one doesn't even make any sense. We have the Fragment, but it's the Urumi who are possessed.

'Bright pinnacle of darkness'? That one could mean anything. And the outcomes are oxymorons, as is 'common difference'...."

"Henry. Stop. I was just quoting from the summary I read of the original passage. I probably left some details out. You aren't going to figure it out in five seconds. Don't try. It's not like it's our fault anyway. The original medium apparently sucked at description."

"But, um...."

"The Society – that's Arunesh and Zitar – has pored over the original text. They are aware of our situation. If they believe that they have any relevant information, they will send it to Malka or myself."

"How?"

"You've seen how," Liza continued, grousing. "Then, of course, whatever they send is usually tight-lipped and annoyingly cryptic."

"Because?"

"Because, security. Henry. It's like I told you: any new info they send out is easier for the Urumi to get their hands on. That decreases its value. Despite all the mystic stuff, it's still like a good, old-fashioned spy game. Anything we get is likely to make sense only to us, and only in a certain context. That means we get to spend most of our time working in the dark. Fun, isn't it?" The felinoid finished her latest complaint.

Henry, noticing that the black-haired girl seemed to be in a talkative mood, attempted to press his luck further.

"So, what exactly is the Society? Where is it? And, uh, who are these people you mentioned?"

"Not now, Henry. It wastes time and we have a crisis situation on our hands. We need to get back to business." The green-eyed were-cat turned to Malka. It quickly became apparent that the Thag had barely registered most of the discussion that followed Liza's recounting of the Prophecy's content.

"Malka, focus!" Liza yelled. Malka looked up to face her.

"Why did he not tell me?" she asked of no one in particular. "So much would have made sense, if he had."

It was true. Looking back on her life in the camp, so many of Husain's decisions and actions made sense in light of Liza's revelation. Why she had been picked as his protégée. Why she had not been allowed to leave the camp. Why he had not blamed her for failure to eliminate her targets. And, why he had called out the entirety of the Thags, when she had only asked for his assistance. It was unnerving. It made the girl feel powerless. Had her entire life been determined by some Prophecy?

No. Malka could remember making certain decisions. They were her own, for better or for worse. Those decisions merely coincided, so far, with the Prophecy's accounting. It helped explain things. But, that was all. However, there was one thing that it could not account for. The question burned in her mind:

"Why did he not tell me?" Husain's former pupil asked again.

"How the hell should I know, Malka?" Liza asked. "Maybe he was trying to be nice? The Prophecy isn't exactly a description of a nice long vacation."

"No. He was not," the Thag replied immediately. She thought of all those instances during which she had confided her fears over her position in the camp to her Master. Now, she felt rather certain. This revelation was what he had refrained from telling her. Yet, Husain could have made things so much clearer.

"I don't know. He ran out of time then?"

"No. He'd have made the time."

"Right, Malka. You can bemoan your past later. But, now that you know what happened wasn't your fault...."

"But it was, Liza," Malka practically yelled, moving her face towards the felinoid's. "I lied to Husain. I told him that the Fragment was in the caravan." She slouched back in the chair. "I'd left the jewel with my sister. I wanted to rescue the three held captive. If I had told him my motives in full, he might not have ordered what he did. The existence of some Prophecy does not absolve me of this. If I had known of it sooner, I might have told him everything. Even if half of them hated me, those people were the closest thing I had to anything. And I...."

"Um, Malka?" It was Henry. "Can I ask you something? Would you have even asked Husain for help attacking this caravan, if you had known what the consequences would be?"

The Thag pondered the question for at least a full moment. There would have been death for some, in either case. But, with the Fragment switched out, its power would have been safe from those who sought it. If she had not asked, the Fragment would at least be safe. The Thags would still be alive.

But the blue-eyed girl wondered now if that really were the case. The Urumi had known – somehow – that Antonia had the diamond. They had known who and where she was. The destruction that Prince Lubomirski and his henchman would have unleashed that night at the Invisible Circus – had she and Stas not been free and able to take steps to stop them – would surely have overwhelmed the Thags in the end. In a wider sense, that outcome was far worse. Still, knowing what the ending of that battle with the caravan would mean, could Malka have brought herself to make the request?

"I do not know," Malka whispered eventually.

Henry sat on the couch directly across from the subcontinental native. He bent forward so that he could look her directly in the eye.

"Maybe," he said, "he was trying to protect you. To make it easier for you to take a step that he knew would be difficult."

Malka shook her head.

"There is nothing in what Liza has said that would provide for that."

"No. But look, Malka. Despite having been around you for months, I still don't know that much about you. I've accepted that. I mean, two months tied up with one of those Urumi things made me realize that this is a lot better. Well, that – and that I have no place else to go, really. But anyway, it seems like you were in that camp for a long time. I gather this guy knew you pretty well."

"He did, better than anyone else in the camp, at least," came the Thag's reply.

"So, then, I don't really think my explanation was so far off. He apparently wanted to help you through this decision. And, he probably knew this Prophecy well. Studied it. That meant he thought his decision was a necessary one. Don't berate yourself for something that had to be. Okay?"

Malka sat up. The girl made a visible effort to square her shoulders. She exhaled a long breath.

"Okay."

Henry offered a curt nod.

"Great. Now that all the worrying, sulking, memory-wallowing, communication breakdowns, random curiosity, and self-remorse mongering have been taken care of, can we *please* concentrate on coming up with a new action plan?" Liza grated.

In that moment, a piece of parchment dropped from thin air onto the writing desk that stood in the room's left corner, next to the windows. Resolutely, Malka moved to pick it up.

As she did so, Liza looked at Henry, briefly smiled, and mouthed the words 'Thank you' at him. In response the boy smirked and retorted, "You know something, Liza? You left something out of that list. You've spent the better part of the past hour criticizing and complaining."

The felinoid smirked, but not in her usual condescending manner.

"I'll whine if I want to. It's comforting."

"Uh, sure whatever. It's not like it can ever be wrong when you do it."

"Stop arguing. Both of you," Malka said as she returned to join them. Both Henry and Liza snickered in response. The Thag sighed.

"Would you like to know what is written here or not?" she queried.

"Okay. All right, let's hear it," her protector responded.

Malka read out loud what was written on the parchment:

> *The Mała Bint lives still at the Entrance.*
> *To the new sphere of stone.*
> *Walk along the noble path.*
> *The patron sets forth from the empty soul.*
> *Of a people self-betrayed.*

> *Now quit whining. All of you. It accomplishes nothing.*
> *-Arunesh*

"Lovely. Tightlipped, annoying, cryptic, and needlessly poetic. What did I tell you?"

"Poetic? It simply makes no sense," the Thag commented.

"Well, I guess you could look on the bright side," Henry offered. "At least the last line is clear."

The felinoid glared at him.

"Okay. You're supposed to be so smart, or widely read, or whatever. You tell us what it means." Liza snatched the paper out of Malka's hand and gave it to him.

"Well," he began, "the second word has a diacritical character. That means that it, and probably the word after it, are in a language other than English. Maybe it refers to this other object the Prophecy mentions."

The green-eyed girl shook her head.

"It doesn't make much sense to refer to an object as living. That part's got to mean something else."

"Hmm…," Henry continued, "without being able to translate this phrase, I can't really tell any more. It might help some of the other references make more sense."

"Henry, is there somewhere you can go in this…," Malka paused, searching for the right word to describe the city, "place, where you could gain access to the resources that would allow you to do so?"

"Well, I guess so. I think they're called public libraries, or something."

"Great. Fine. Go there," Liza said, clearly frustrated at this newest delay.

Henry did not move.

"I mean now. Henry," the felinoid clarified.

"Um, Liza…."

The irate protector sighed in exasperation again. "What is it?" She said the words through almost clenched teeth.

"Um … it's the middle of the night? I doubt it's going to be open…," Henry looked at the gilded clock on the mantelpiece, "at three in the morning."

"Oh, right," the green-eyed girl said, slightly embarrassed. "Sometimes I forget about opening hours. Being who I am and in this line of work, I don't have much use for them." She paused again before continuing. Again, she allowed her frustration to show:

"So, we'll just have to wait around until this damn place opens. Amazing. How much longer are we going to drag out this mission? It's already gone on way too long. Then I can finally get away from you people!" The felinoid turned on her heel and headed into one of the bedrooms, slamming the door behind her.

"Thanks, Liza. Nice to know that you're enjoying our company," the blue-eyed boy called after her.

"Oh, shut up, Henry!" could be heard, muffling through the wall.

He turned to Malka. A slightly confused expression was on his face.

"What got into her, there?" he asked.

"Believe me, I do not have the vaguest idea," the Thag responded, shaking her head.

"I can volunteer to take the couch, if you'd like," Henry broke the silence.

At this, Malka smiled slightly, if only for a second. "It is all right, Henry. Where I grew up, we used sleeping mats. I am content to sleep here, on the carpeting."

"Oh, I see. I didn't know that. If you'd rather take the floor, then okay." He turned to head into the other bedroom. Malka watched him go. He was from another part of the world and a completely different background. Yet, he had been raised in an isolated environment, different from those around him, just as she had been. The Thag did not know for certain if that meant he could understand her past. After leaving the camp, she had been reluctant to tell anyone – even Stanley – all of her story, for fear it would be dismissed, providing more confirmation that she truly belonged nowhere. Instead, she guarded her identity with an automatic defensiveness.

Yet it was only now, as she dwelled on Husain's last piece of advice to her, that the girl began to suspect. Although he had not told her all of it, her Master may have been trying to impart a hint from the Prophecy that would help her to find allies – ones she could trust – on the journey that she had unwittingly been about to embark upon. Common suffering, even if unshared.

"Henry. Wait a moment," she called just as he reached the threshold. He turned.

"Uh, yeah?"

"You have recently said that you know little about me. But I have heard enough of your upbringing to suspect that we may have shared similar pasts. I do not expect you to say anything about your own. But, I hope that understanding my past will help to clarify the task with which I am now faced. For both of us."

"Sure. I never really got why you always used to get so ticked off whenever I asked who you were back in the desert."

"I don't know if you will, Henry. For me, that admission is difficult. But the last lesson that my Master imparted to me suggests that in some cases it may be necessary. I cannot ignore that advice."

Henry smirked. "You came all the way across the country to rescue me. And it's still harder for you to tell me about your past."

Malka nodded, with a small ironic laugh. "Identity can be a fragile thing. It's difficult to trust others with it."

The camp-born boy nodded knowingly. He took a seat directly across from Malka. Taking a deep breath, the half-breed Thag began.

Ten

Henry returned from the library. It had been a harrowing morning and early afternoon. Shortly after the sun rose, Liza had emerged from the chamber she'd appropriated the night before. Malka and the boy remained seated in the living room, still engaged in conversation. Over the course of the remaining hours of darkness, Malka had told him much of her experiences growing up in a camp of fanatical thieves. Henry had listened, volunteering some similarities and differences regarding his own past. He had also seemed not to judge her beliefs, actions or fears.

As Liza entered the living room, it was clear that her consternation had not abetted.

"Great. I leave the two of you to your own devices. Grab a few hours of shuteye. Then I find that you've decided to use the time to pull some kind of all-nighter."

She paused, looking at the ceiling.

"If you haven't noticed, the sun's up now. Whatever you were doing, it's time to get down to business. Henry, get lost. I'll try to get Malka to focus, for once, on our situation."

"Great. Fine," the brown-haired youth replied. He headed for the suite's front door. "It's not like demeaning her will get you anywhere. She's too used to it," he murmured by way of departure.

Having been born and raised in a mining camp, it had been Henry's first trip to a library, not counting his own private collection of books. He had convinced his parents to bring them to him when they went on trading or provisioning runs.

First, he'd needed to find the place. Henry had attempted to ask directions. However, most greeted his enquiries with suspicion or contempt. They were almost as derisive with the boy as most in his camp had been. He'd ended up stumbling upon the building by accident. Noticing the sign that declared it to be one of this metropolis's main libraries – the one that bore the name Astor – he'd entered.

Another dilemma presented itself inside. Having never entered such an institution before, Henry found that he had no idea of how to find the relevant materials. The docents had eventually showed him how to use the library's cataloguing system. But, they had done so in a manner that clearly suggested they thought he was an idiot. As far as they seemed concerned, he should have known before walking in the door. Henry had thought that bit odd. It was their job to explain how things worked here, as far as he knew. Yet, they acted as if he were some kind of charity case from the countryside. Judging by their attitude, they were doing him a favor simply by explaining how the system worked.

Feeling somewhat harrowed by the experience, the boy was finally able to research possible meanings of the message sent by the Society. As he did so, it was with the distinct idea that he was developing very few positive impressions about this place, or the people who inhabited it.

Now, Henry reentered the suite at the Fifth Avenue Hotel. He was thankful that he could once again isolate himself with those he had once thought of only as his kidnappers. The young man had been gone for just a few hours. To him, it had seemed much longer. He also feared that his investigation had hit a wall rather quickly.

Henry noticed that Malka and Liza remained seated in the living room. Both had frustrated looks on their faces. It was evident that they were arguing.

"I still don't get why you're insisting that this changes nothing, Malka. It's moronically wishful thinking."

"No, Liza. It is not. Even if the Urumi can take me into their Order, there remains nothing that we can do about that fact. You're insisting that there must be something we can do and that I simply refuse to find it. That's what is fanciful!"

"Um, ladies?" Both of them looked up at Henry. "Any progress?"

"No," they both spoke at once and pointed at the other. "She is being stubborn."

"Well, I just thought I'd let you know: I have some information. But, I am not sure how much it's going to help."

"Lovely," Liza replied. "All right. Out with it."

Henry took in a breath, beginning his explanation.

"The special letter I told you about? It's Polish." He paused for a moment, deciding to provide more context. "Well, it's not only Polish, actually. It's also used in Silesian, Kashubian, and Sorbian, which are some of the other...."

"Whoa, whoa, whoa. Too much information. Cut the irrelevant stuff," the felinoid snapped, looking at him sharply. "Point being?"

Henry sighed.

"The word it's used in is Polish. It means 'little,' or 'small.'"

"And the other one?" Malka asked.

The camp-raised boy sighed again.

"Well, that's where I started to run into problems. The other word doesn't show up in any Polish-to-English dictionaries, which means it's either some informal usage or that it's in another language. But, without any special characters, there is no way to narrow down the list of possible languages."

"Great. So we're nowhere." Liza allowed herself to slouch lazily back onto the couch where she sat.

"Hold on, Liza. Since when did you give up so easily? The fact that at least one of the words is in Polish allowed me to figure out some other things about what we were sent. Maybe. At least some of the other phrases make sense with reference to Poland. Or, what was once Poland, anyway.

"For instance, I did some general reading about Poland, hoping to stumble across some clues. It appears that the country broke up, not only because of neighboring foreign powers, but also because the Polish nobility was complicit with them. They wanted to keep their own feudal power. So, they sold their own country down the river."

"'A people self-betrayed,'" Malka interpreted in a breathy voice.

"Uh, right. And the reference to a patron? I'm not sure what this tells us, but that could be a reference to one of Poland's patron saints."

"Henry, *does* this – any of it – actually tell us anything?" Liza challenged.

The brown-haired youth shook his head.

"I don't know." He sighed in frustration. "Frankly, I sort of wonder if this Society of yours sent us the right note. If we're supposed to be searching for some object connected with the Fragment, why is this note about a living thing connected with a former country in Eastern Europe? Even once you take information security into account, the directions to this other object's location were discovered in India. So why don't we have clues centering on India? Clues made easy for us – and only us – to interpret?"

He let the questions hang in the air for a moment. Both young women met them only with silence. Allowing his point to sink in, the blue-eyed boy continued:

"You said that what the Society sends sometimes only makes sense to certain people in certain contexts. Besides the word I can't translate? A lot of the clues are vague. For instance, 'sphere of stone' and the reference to an 'empty soul'? Maybe they would make sense to someone who knows more about Poland. But, without context-specific knowledge, they are too abstract for me to know where to look. I'm sorry. I wish I could tell you more, but it could be that this note is, you know, not even meant for us." The mining camp scholar paused, as if questioning his own competence. "I just don't know." He sat, resting his hands in his lap.

After a few seconds, Liza broke the silence. She appealed to the divine, followed it with a choice expletive, and then made a reference to damnation. The outburst surprised the two others in the room. They had known the felinoid to be somewhat direct. But, neither had ever heard the black-haired girl be downright profane.

"Just amazing," the green-eyed girl continued. Standing, she raised both arms out to her sides. "Finally. We get some intelligence. It turns out the Society was just screwing the pooch! How the hell am I supposed to work like this!" She turned away from the other two youths. The milk-skinned figure moved to stare out the window, looking out over the streets of Manhattan.

Again, silence. After a moment, Henry offered, tentatively: "Um, Liza? Are you okay?"

When the felinoid responded, it was in a low, cold voice.

"Henry, can we please discuss anything other than how I am personally doing right now." It was a demand. Not a question.

"O-kay," the boy replied, drawing out the word. After a moment, Malka spoke up. There was a questioning element in her voice.

"Henry, I agree that this note may not have been intended for us. However, I do not believe that it was sent to us by mistake. This happened once before. It was the first time I received one of these notes, as I told you, back in Madras in the bungalow."

"Right," Henry prompted. She had told the youth of her adventures on the subcontinent with Stanley and Mungo, yet Malka recounted part of the tale again.

"It stated that the friend, for whom Stanley was searching, was in Pondicherry. At first, I thought it was sent to me by mistake. So, taking Mungo, I went to follow Stanley as he moved about Madras. I had been intending to see if he received a duplicate note as well, though my reconnaissance was cut short."

"Uh-huh. That was the first time you encountered an Urumi."

"Yes. Mungo and I repaired to his family's beach home directly after. We decided it was best. I had been fairly certain that the Urumi no longer watched us. Mungo injured it during the battle. I saw the blood."

"Okay."

"I don't know from where. But, eventually Stanley returned with a similar note. It had the same emblem. A snake eating its own tail. It stated that my sister was also in Pondicherry. We were able to exchange the information, relevant to each other, without the knowledge of the dark warriors. I suspect that this may function as another layer of secrecy. If the Urumi were watching us, they saw us receive a note with seemingly irrelevant information. It looks like a mistake. But, that allows us to exchange information at a later time, without their knowledge."

"Okay. So it's a deeper tactic on the part of the Society. I get it. But, Malka, so what?"

"I cannot be certain," the Thag replied slowly, her mental wheels still turning. "However, the fact that the message clearly refers to this 'Poland' and was sent to us may carry information in itself."

It was here that the felinoid cut in.

"Oh, please, Malka. This is ridiculous. I've been working for the Society for over a decade. I have never seen them do this type of thing. Why would they? It's bad encryption strategy, for one. How in the name of perdition are they really going to expect us to get it to the right person with any level of certainty? Yeah. It makes it hard for the Urumi. But, you want encrypted messages to be immediately decipherable to the intended recipient. This is just convoluted. Needlessly." She continued to stare out the window.

"I am not so sure, Liza," Malka countered. "As I have said, I know Arunesh and Zitar have already done this once. They sent a note, apparently to the incorrect person, when the Urumi are likely to be watching...."

"Are they?" Henry asked, interrupting.

"No," the Thag replied. "I would sense it, if I were being observed covertly. I am not. And that is strange. It would be reasonable to assume that they would be doing so, at the very least. Currently, they have relatively certain knowledge of our location."

"You can say that again," Liza graveled softly, shaking her head. "This entire mission is going down the tubes."

Ignoring the felinoid's assessment, Malka and Henry continued their conversation.

"So," the blue-eyed boy reasoned, "it would be overwhelmingly reasonable to assume they would be near at this point. But, for some reason they aren't." He nodded as if realizing something. "That gives us an added advantage."

"Correct," Malka confirmed. "We can openly figure out for whom the message was intended, without having to puzzle it out only in our own minds."

The felinoid turned from the window. She spoke in a low voice that dripped with condescending sarcasm. "So what do you want us to do, Malka? Go walking around the Polish neighborhood in Greenpoint asking people if they understand this message?"

Malka turned to regard the felinoid. The Thag did not completely understand the statement, but its meaning was sufficiently clear for her to have become truly annoyed herself. The blue-eyed girl responded with uncharacteristic directness.

"*Think* about it, Liza. Arunesh did not send this note to anyone. He sent it to *us*. How many people do we three know that have some connection with Poland?"

Both Liza and Henry shook their heads. After a beat, Henry spoke.

"Wait a minute. You know someone, Malka. You told me last night. One of the two you were assigned to eliminate. You said he was Polish. Except that he wasn't from there. You thought that was interesting."

"Yes. Stanley."

"Where is he now?"

"He remains in Madras, I assume."

"So we're going to have to go all the way back to Madras?" There was a nervous anticipation in Henry's voice.

At this, Liza moved toward the windowsill. She placed her elbows on it. Then she bent forward slightly, resting her forehead on both of her wrists. The black-haired girl let out a bovine groan. It was followed by another disbelieving expletive.

"You people!" she continued. "Just … send a telegram!"

"How do we do that?" Henry asked.

Liza turned, attempting to calm down. "You go to a post office," she explained with an annoyed air.

"A what?" asked Malka.

Liza sighed, groaning again. "Do I really need to tell you how scary it is that I am just supposed to be here as the added security, and I'm the one that has to tell you *idiots* how to do even basic things?" Another shocked silence fell over the room.

"Um, Liza," Henry eventually ventured. "I get that to you, these things might seem like basic common sense. You have more experience than we do in these kinds of places. But, you need to remember: to us, with our backgrounds, it's all very new. We might still need some explanation from time to time."

At this, Liza's mien softened suddenly. She allowed a slight smile to play across her mouth.

"Very well. Malka, you have Stanley's contact information?"

"Yes. It remains in my satchel." The Polish boy had given it to her just before her departure from Madras. The Thag had been pretending to be her sister. She had accepted it. Yet, she was not quite sure what Stanley expected her to do with the information. The blue-eyed girl heard Liza continue:

"You're at a very upscale hotel. So, actually, you can just go to the concierge...."

"The what?" Malka interrupted.

Liza sighed.

"Just look for the sign in the lobby. You can read, can't you?" This time, it was clear that the felinoid meant the insult as a joke, intended to soften the accusation of her recent outburst. "Give Stanley's address to the man behind the desk. Tell him the message you want to send. Keep it short. Got it?"

Both camp-raised youths got up and headed for the suite's front door. As they exited into the corridor, Malka turned to Henry.

"Is it only my perception, or is Liza becoming progressively more...."

Henry finished for her, choosing the exact word that she was about to utter: "Erratic." He nodded. "Do you have any clue what's...."

"Bothering her?"

As they entered the intricate, brass-caged elevator, the Thag shook her head. Outwardly, it was a gesture meant in response to Henry. Inwardly, though, it reflected her final acknowledgement of another painful necessity that their overnight conversation had forced the Thag to countenance.

Eleven

A response to their dispatch came after only a few hours. Malka had strongly suspected that Stanley had seen through her disguise, back on the beach in Madras. Yet, she had elected to keep up the subterfuge that the Thag was her sister, Antonia. The message from Malka had been short. It had been quickly worked out between her and Henry in the lift, on the way to the lobby. Its content included a simple statement: She – Antonia – had been sent by her order to a diocese in America. There she worked mainly with Polish immigrants. The update was followed by a query as to what the phrase *Mała Bint* meant.

However, the camp-raised girl had been surprised to discover that the response she had received was not from the youth whose country no longer had a place on the map:

```
       Antonia, good to hear from you.
   Stanley moved to Switzerland half-year ago.
     Attends St. Nicholas School, Fribourg.
       Did not send contact details.
          Still miss your sister.
               – Mungo.
```

Malka could have reported some mild amusement at the second-to-last line of the message. Henry had snorted upon reading it. The Thag had told him the night before how, during their adventures, Mungo had become naively infatuated with her, or at least with her mystique. As such, he'd held an automatically positive opinion of the Thag. That had been why Malka felt that she could entrust him with some partially fabricated information, carefully selected facts regarding her background. And, the admission that she'd tried to kill him.

Henry had nodded in understanding. It now appeared that the red-haired Brit still suffered some of what the American-born youth had termed 'puppy love' when he'd heard Malka describe it. The blue-eyed girl's decision not to reveal her true identity, in her communiqué to India, had been correct.

However, the lines above it were cause for much greater trepidation. There was no contact information. It would not be possible to send another dispatch. Further, the Thag decided that now was the time to keep moving. If they left when the Urumi were not watching them, it would be harder for the Shadow Warriors to find them again. Keeping on the move at this point was the best way to exploit their enemy's apparent tactical blunder.

At the same time, Malka's conversation with Henry had made the Thag realize something about her own motivations. Her initial decision to take the boy with her arose not only from a need for secrecy or security; it flowed from a desire to do for him what Husain had done for her. At some point, she'd thought it might ease her remorse, and re-create some wisp of the semblance of community that she had lost. But Malka now realized that, unlike herself, Henry had some basic knowledge of how his land functioned. It seemed that those who had been with him in his camp were at least of the same culture. He knew how banks worked. He'd seemed to at least have heard of a post office. She did not want to require her erstwhile captive to leave behind the culture where he'd been raised.

Malka trusted the brown-haired boy. She knew that he was unlikely to reveal their actions to law enforcement, if set loose. She was certain that he was intelligent enough to realize that if he reported them, he would likely implicate himself in their larcenous activities. Further, Henry's kidnapping by the Urumi demonstrated that he was – as Liza had pointed out from the beginning – a liability. Getting the boy back had not only put her core mission at risk; it also put Henry in harm's way. That was something for which the Thag no longer wished to take responsibility. Mostly, though, Malka had come to identify with the camp-raised boy. The Thag did not wish to rob him of a chance to live a normal life with his own people. It was something she'd never completely had. That chance was the least the half-breed felt she could give her newfound friend and ally.

Shoulder to shoulder with Henry, the Master of the Thags marched towards her suite's front door. Her decision was made. Opening it, she entered. The blue-eyed girl made her announcement:

"Liza and I shall depart for Switzerland immediately."

"What!" Henry and Liza yelped in unison. It was for different reasons:

"Switzerland? Why the hell do we have to go all the way there?" Liza moaned.

"Wha-what about me?" The brown-haired boy's voice carried more than a hint of injury.

Malka began with the easier of the two queries.

"Apparently the person who we seek moved there. To a place called Fribourg. He did not leave the information we need to send another dispatch. Besides, it makes more sense to remain on the move when the Urumi are not watching us, for whichever reason that may be."

"Strategically sound." The felinoid appeared satisfied. Then, she added, "Even if it means I get to enjoy the pleasure of your company, as we drag our behinds across yet another ocean. Yay."

Ignoring the felinoid's commentary, the Thag turned to Henry.

"Henry, you have been of assistance to us. I will not enjoy leaving you behind."

"Then don't," he interjected. "I can't believe that you would come all this way to rescue me. You risked the Fragment to collect me from those Urumi things. Now you're just going to strand me here?"

"The longer you stay with us, the more chance that harm will come to you. Now that we have gotten to know something of each other, I do not want to see that happen."

"Who said it was your decision?" he snapped, hurt.

"She's paying for the boat tickets, Henry. That's what," Liza interjected, looking away from the discussion.

"Yeah. With money I helped to steal. It's not fair."

"Only so you could escape," the felinoid reminded him, sighing before drawing another breath. "And you might not want to call attention to the fact that you engineered a major crime, while trying to lecture someone on the concept of what's fair. It undermines your credibility, if only slightly."

Henry appeared undeterred: "I only tried because...." Malka cut him off.

"That is not the point. Mostly, Henry, I am doing this because, despite exactly where you were raised, you seem, to me, somewhat like one of these people. Liza and I are doing what we must. But that is not the case for you. I want you to have the chance to live a normal life with your own people. If you come with us, I fear you may lose that opportunity."

"My own people?" he responded, not angrily. "Huh, I guess it might seem that way, from your perspective. But, have you considered how things look from mine and, you know, from theirs? In the camp, everyone thought I was worthless for having interests other than endlessly chasing some dream of striking it rich by digging around in the dirt for most of my waking hours. From what you have told me of your own past, Malka, many things made you different from the others in your camp. Ethnicity, or birthright, isn't enough to create belonging. Even though we look the same, speak the same language and have the same accent, those I have met from outside the mining settlement seem little different from those inside. If being like that is what's required to have a 'normal life' – whatever that means – with these people, they can have it. Don't say that you are doing this for me. It's nothing I want."

The Thag smiled sadly, shaking her head.

"You may feel that way now, Henry, but for all of my life I have been ostracized. Never accepted, not only because I had my own thoughts, but also because my origins and experiences were fundamentally different from all others in the community where I found myself living." She shook her head, starting to pace. "I fear that the same may happen to you, if you come with us. Here, at least, you have some basis for commonality. I am worried for you, Henry. I'm concerned about what will happen if you come with us: that you will find yourself in a world that will always treat you as the outsider, but that you yourself will have become too changed to ever return to the life that was once yours, if only partially."

"Like I said, maybe on the surface it looks that way to you," the brown-haired youth responded. He paused, as if desperately searching for the most cogent argument. Finally, shaking his head in anguish, he continued, "It's hard to explain, but I don't *feel* like a part of these people: the way they do things, what they think is important. We don't share that; we never have. I may have the prospect of being treated like one of them because of where I was born or how I look. But don't you get it? That doesn't always matter." He emphasized each of the last three words. "It doesn't matter because I will always feel like the outsider anyway, no matter what. You and Liza are the closest thing to a family that I have."

It was a serious exchange. Perhaps because of that, Liza felt the need to weigh in: "I hate to break it to you, Henry. But basing your conclusion on one mining camp, a backwater western town, and one city that's world-famous for its bad attitude? Maybe not the best way of making that determination."

The blue-eyed boy looked nervously back and forth between the other two in the room. Malka turned to look him directly in the eye, fixing him with a sympathetic look.

"I do not want to leave you here. But, Liza is right. I was the outsider in a culture of one camp. That is not the case with you. You are so fortunate. It's for the best. I'm sorry."

Henry felt as if the floor had just been ripped out from under him, even though he continued to stand, shocked, in the same place. When Malka had kidnapped him, months ago, the brown-haired youth would have found it hardly believable that he would so desperately want to go with these same two. His mind raced, looking for some way of convincing them. Malka and Liza moved to pack their things. The felinoid rang for the porters to collect their luggage chests.

Then, Malka walked up to Henry. She told him that she would leave a third of their combined haul of cash with him. It wasn't like she really cared about keeping it herself. For the Thag, the money was only a means to an end. But, Henry had played a key part in that quest, whatever his motives at the time. He deserved a share of the trophies, such as they were.

He'd nodded numbly upon hearing the news. Logically, his mind told him to cheer. It was enough to allow the camp-raised youth to live in relative comfort for the rest of his life. Yet, he just stood there, alone. An island of one.

After a few more moments, Henry vaguely perceived that the porters had taken the trunks, containing Malka's belongings, out into the hallway. Malka and her protector moved towards the suite's front door.

Stopping in front of him, the Thag paused. The girl looked at him sadly for a moment. Then, she threw her arms around him. Henry returned the gesture. He could hear that her breathing was ragged.

Eventually, Malka stood back. "Be well, Henry." The girl's face remained dry, but her eyes were watered. Her voice cracked as she said the words.

"Be well. I hope that your quest meets with victory," the boy managed. Malka turned and moved towards the doorway.

"Bye, Henry. Who knows? Maybe we'll run into each other again someday," Liza said simply.

A sad smile of resignation played across his features.

"I'd like that, Liza."

The two young women headed for the doorway. Henry's brain reeled with the sudden shock of what had happened. He'd finally found some semblance of belonging. Now, those with whom he had found it were leaving him behind. Their minds were made up. He could not think of any way to convince them.

Then, he hit on something. It had nothing to do with who he was. Nor was it a threat to reveal their actions and mission; he knew that wouldn't work. It did not take issue with the reasons they had given for continuing on without him. Henry did not know if his last-ditch effort would meet with the desired effect. The boy tried, anyway:

"Et parlez-vous Français?" Liza had been about to shut the door when she heard Henry call after them. The felinoid and camp-raised girl turned to look at him from the hallway.

"Parlez-vous Français?" Henry repeated. "It means, 'Do you speak French?'" A short silence. "I take it from your expressions that the answer is 'non'? If you're going to Switzerland, I assume via France or Belgium, you could find it very helpful to have a French-speaker along for the journey."

Liza looked simply perplexed.

"Um … since when did you become fluent in French?" She shook her head slightly, moving it backwards in noncomprehension as she drawled out the words.

"Back in the camp. I taught myself from textbooks, novels and magazines."

"Okay, that's *totally* normal. How many languages, pray tell, did you learn there?"

"French, Latin. Some German and then a bit of this language that they speak in the Austro-Hungarian region of Carniola. It was more interesting than digging around in the dirt. It also means that having me along could be very useful."

The Thag remained skeptical.

"I am not sure, Henry. I do not want to take you away from any remaining connections to your people."

"Malka, we've been over this…."

It was Liza who interrupted him.

"All right. This is a pathetic excuse for a last-minute gambit. Should have just slammed the door on my way out," the felinoid yelled at him. "I mean, the desperation alone is enough to make me go running in the opposite direction." Henry's heart sank as the felinoid snidely drawled the words. Not only had his plan not worked; it now looked as if he would also part ways with them on rocky terms.

Liza paused for a moment. Then, she turned to Malka and continued, a slight smile on her face: "I say he comes with us, Malka. We need him."

"But Liza, how important is knowledge of languages really going to be?" She paused, looking for the appropriate phrase. The idea that language would be so important seemed odd to her. In India and America, she had not encountered problems using English to communicate, even when other regional languages existed. Could things really be so different where she was going?

"Oh, believe me, Malka. You have no idea." Liza paused for effect. "Henry's right. We should bring him with us."

Malka regarded Henry for a moment. Then, she made a decision.

"Very well. He comes." The Thag felt a surge of relief as she spoke the words, yet she kept her voice as neutral as possible. As if the entire discussion had never occurred, the blue-eyed girl turned and walked down the hallway. The felinoid followed suit.

With his own immense sigh of relief, Henry joined them. A tight smile was on his face. The trio – for that was what they now were – made their way down into the lobby. Striding shoulder to shoulder, they exited the front doors of the Fifth Avenue Hotel and made their way into a waiting carriage.

"Where to?" the driver asked as they entered.

"Passenger port. Make it fast," Liza purred in response. The conveyance began to move with a sense of urgency.

With that, the three of them set out for the shores of Europe.

Twelve

The passage that led to the courtyard of St. Nicholas School's dormitory was dank. Stas entered, not wishing to linger. The tree within the outdoor enclosure had long since sprouted leaves, shading most of the open-air space. The weather in Fribourg had finally warmed up to some halfway comfortable level. The courtyard would have been somewhat pleasant. Yet, despite the fact that it was mid-May, it had refused to stop raining. In practice, this made the courtyard damp and buggy. Stas hurried through it.

Moving to the dormitory's only entrance, Stas stopped briefly at the front desk. He asked after his room key and mail. The key was not there and he had none. While the first of the two facts meant that Jurgen was likely present in their dorm room, Stas was relieved at the latter. After learning of Nell's passing, he, if pressed, would have had to confess a bizarre fear. Namely, that more unthinkable news would be waiting for him every time he received an envelope. Worse, most of the messages he did receive were sent by his father, whose letters continually urged Stas to accept his loss and move on. Nell's protector found that more than difficult. When not studying, his mind invariably turned to memories of their adventures together. The fact that there would never be new ones gave the once-fond recollections a melancholy tint. Stas was alone. He felt miserable. He did not think that he would ever truly get over Nell's death. What was more, he found that he did not want to. Somehow, that would seem like betraying her.

Swatting a few lazily airborne flies out of the air, he ascended the dorm's creaky wooden staircase to the building's highest floor. Turning, Stas arrived at his quarters, opened the door, and entered. He almost ran headlong into Jurgen.

The Swiss German was pacing back and forth in the small floor space of the dorm room. His manner evidenced a lack of the reserve that he'd usually held. The Slav edged around him and moved to his desk, placing his textbooks and notes on top. His roommate continued to pace in an agitated manner. In his right hand, Stas noticed that he held a letter, half crumpled in his palm. In his other hand he held a silver and black walking stick, though Stas had never seen it before. He couldn't say where the formally dressed boy had gotten it.

The pacing continued, until it began to grate on Stas's already frayed nerves. He attempted to force himself to focus on course material that he already knew quite well. It was the middle of the spring exam session. The Slav had just returned from the school's library, where he had been reviewing his books. Of course, he already knew their contents backwards and forwards. But, the exercise of study helped to take his mind off of reality. Jurgen's pacing interfered with that.

"Jurgen," Stas said finally. "Maybe you should sit down and study. We have our mathematics examination tomorrow at noon."

The pacing continued.

"I do not think that I can study at a time like this," the thick-framed Swiss German murmured.

Nell's protector had no idea what Jurgen was talking about, but he was instinctively annoyed by the comment. After enduring the loss of Nell, he had numbly carried on. Stas did not know what was bothering Jurgen. But – whatever it was – it seemed self-evident to Stas that the apprehension under which his roommate was laboring could not be worse. The young Tarkowski framed his response:

"If I was able to keep up my studies after what happened to me, then you most certainly can. Now, can you quit your pacing, please?"

Jurgen continued to patter back and forth.

"Yes, Stanislas. I am aware of that. Study is all you do anymore. Heaven forbid that others have problems and deal with them in their own manner." Coming from the Swiss German, it was a shockingly direct statement. Since their conversation on the hill, the two youths had gone out of their way to give each other a wide berth, despite sharing cramped living quarters. Their communications had been limited to basic necessity.

"Whatever has happened, it cannot possibly be as bad as you think it is. For all your criticisms of my behavior, I never acted like this after I found out...," Stas's voice trailed off.

"Yes you did," Jurgen mumbled. "And then you kept me up endless nights with the lamp burning, so that you could sit hunched over your books." He continued pacing. "It is as bad as I think it is, possibly worse. There is nothing I can do about it. Absolutely nothing."

Stas was growing impatient. Jurgen's demeanor reminded him of his own powerlessness to defend Nell from her fate in the end. He had gotten her through the African wilderness and rescued her in India, though it galled him that he still could not explain from exactly what. Was it all so that she could perish from exposure in faraway England? It was senseless. That was the cruel turn of fate, which proved almost too much for Stas to bear.

"Then lie down, stare at the ceiling and worry. Quietly." The Slav's voice was soft but it carried an edge.

"I can't do that. Not at this moment. I have received very disturbing news." Moving toward Stas, Jurgen extended his right arm – the one with the letter – momentarily. He stopped, briefly, as if to emphasize his point. Then, he went back to pacing. Explaining as he did so:

"My father never told me about this. As one of the more prominent families in Steckborn, we ran one of the major shipping companies that sends cargo up and down the Rhine. He had been teaching me to run the business but he keeps most high-level things to himself. Before I left for Fribourg, I wondered why we were charging lower and lower rates, sometimes lower than cost. My father always avoided telling me. Pressing harder would have been improper behavior." The native Swiss shook his head as he paced.

"Presently, I have found out the reason. He was attempting to take over one of our major rivals. I know the person who ran it. He and my father went to school together. They could not stand each other. Apparently, my father had been about to successfully take over his company, which my father had driven to the brink of ruin by charging below-market prices for our services.

"My father only found out when he came for the corporate seals. Instead of conceding, his rival had scuttled all of his ships." Jurgen stopped pacing, driving the end of his walking stick into the floor as if to make a point.

Stas listened to the animated explanation with a bland expression on his face. His closest friend was dead, and Jurgen was all excited over a few sunken boats?

"So your father had a bad run of business. That's what you're so worked up about? It's not that bad," Stas replied, not mentioning how much he felt that the import, which Jurgen apparently placed on his father's business rivalries, demeaned the loss of his closest friend.

"It's bad. In fact, it could not be worse. The main point of the letter, which I have just received, concerns how my father was able to afford driving his competition into bankruptcy. What started out as a business move became intensely personal. That was why my father did not want to tell me or my family. It was not proper behavior for a civilized, Western gentleman. It had been going on for years, apparently. To finance undercutting our competitor, he'd started taking out loans to fund our company's operations. When our business had reached its borrowing limit, he took them out personally.

"If he had been able to liquidate the other firm's tangible assets, he would have been able to pay them off and gain a sizable market share. But that is no longer an option. We owe over half a million francs in debt. Presently, there's no way that my family can repay it."

Jurgen began pacing again.

"It is that bad, Stanislas. If we cannot get the money within a month and a half, come July my entire family and I will be reduced to utter destitution."

"I see," Stas replied in a monotone voice.

"'I see?' That is all you have to say?"

"Yes."

"Then, I don't think you do see, Stanislas. If this happens – which it is going to – the Fischers will go from being one of Steckborn's most prominent families – one with an unassailable reputation – to being pariahs. Out on the street. I don't even know if I can live that kind of life. Begging like some stupid gypsy." He spat the words with contempt.

"You will make yourself capable, Jurgen. If your people are as great as they say they are," said the Slav, coldly. Jurgen stopped again, looking at Stas. His head was cocked to one side, as if not quite believing that he had heard his roommate correctly.

"I do not see how I can make myself capable of living that way, when I have no knowledge of how to do so. I just don't know what I'm going to do." The imperiled Swiss citizen started pacing again. "Or, if there's anything I can do."

"If there isn't, then why don't you sit down and try to focus on your studies."

"Stanislas, I cannot. What would be the point? There is no way that I will be able to continue with my education after this."

"Very well, then. I'm going back to the library."

Stas skirted around his roommate. He departed, leaving Jurgen to his anxious pacing.

Thirteen

The loud bang of a heavy object, hitting the wall and then smashing to the ground, reverberated through the multi-roomed, first-class cabin on the S.S. Rhynland. It had been about a week since the trio of Malka, Henry, and Liza had embarked on the ship. Each of them held a ticket for Antwerp, the ship's final destination.

The carriage had left them off at a ferry port. There, they had needed to take a boat to an island in the middle of the bay. Malka had done an exemplary job of purchasing tickets from the offices of the Red Star Line. The Thag had managed to give no hint of how new she was to these sorts of things.

Thankfully, the status of first-class passengers freed the group from most types of health and customs checks. They had not been required to produce passports, which, of course, none of them had.

Liza had snidely commented that the Thag was at least "making progress with progress" after they'd left the ticket office. Malka had taken umbrage; Henry had judged it to be an unnecessary dig. The Thag claimed Liza had implied that the way of life she had known for most of her childhood was essentially backward. Instead of denying this, the felinoid had scoffed and mentioned that she "wasn't *implying* anything."

They had boarded the ship and been shown to their suite: a spacious parlor and three separate sleeping chambers. Each had its own bathroom. They had kept mostly to themselves. Malka had reasoned – as she had on her voyage to America – the fewer people who saw them parading about, the better. Even remaining within their spacious accommodations, it would have been possible for the trio to have avoided each other almost completely.

However, Liza had insisted on attempting to come up with yet more contingency plans. What if the Urumi found them, or suspected they were headed for a meeting with this Stanley person? The felinoid had demanded solutions to such scenarios from Malka, multiple times. Yet, over the past week, their discussions had descended quickly into arguments or just plain petty bickering. Mostly, it was because the black-haired girl seemed to go out of her way to provoke Malka. Henry couldn't figure out why. Thus far, he'd tried to stay out of the arguments.

Just now, another tiff had gone into full swing. Liza had – yet again – been deriding Malka's parentage, competence, and commitment to protecting the article with which she had been entrusted. Henry knew this was one way to both unsettle the Thag and get under her skin. Remembering what her Master had told her about being provoked to violence, Malka gave consistently metered replies.

Then, Henry had heard the felinoid scream. "That's it! I've had it with you!" Immediately after, the sound of something hitting the wall could be heard, followed by a loud crash. This was the first time that one of their arguments had deteriorated into the two young women resorting to the hurling of heavy objects. The camp-raised boy decided that it was time to intercede. He got up from the bed in his private stateroom, where he had been reading about Polish history, and padded into the living room.

Henry's blue eyes witnessed a tense scene. The Thag crouched in a defensive posture. Her knife was held in her right hand. It was pointed at the felinoid. Malka stood, ready to strike. Liza remained crouched in a fighting stance as well. Her hands were raised as if ready for fisticuffs. Next to them, a coffee table was upended on the floor. The remains of the two-faced clock that had been resting on it lay against the wall behind Malka. As part of her latest outburst, Liza apparently had thrown it at the darker-skinned girl.

"Cut it out! Both of you!" Henry yelled immediately upon taking in the scene. He saw Malka's hand tense around her blade. Immediately, the blue-eyed boy realized that his words could have been chosen more carefully. Moving quickly, he interposed himself between the conflicting parties. He stood where the coffee table had been. Liza seemed to take notice of his presence for the first time.

"Hey, Henry. Nice to see you're off your dead backside for a change." Henry had heard similar things all of his earlier life in the mining camp, from his parents and from others. He was not amused. When he spoke next, there was a certain authority in his voice.

"Sit. Down. Both of you. Malka, put away the knife." The Thag remained unmoved.

"Henry. Get out of the way. You must have heard what she was accusing me of. She tried to attack me."

"I was trying to knock some sense into your head!" the felinoid yelled.

"No, you weren't," Henry snapped. "I've had the distinct pleasure of hearing all of your discussions since we got on this boat. You were just yelling at Malka for not having a detailed contingency plan for every 'what if' question you manage to come up with. Half of the things you're talking about aren't even in our control."

"She pulled a knife!"

"Only after she attempted to hit me with a clock," the Thag countered.

"Liza, sit down," Henry said.

"You seriously want me to sit down when...."

"Sit down, so she can put away the knife." Henry seemed beyond annoyed.

The green-eyed girl lowered her hands and sat on the dark green divan behind her. The Thag followed suit, lowering herself into a chair of the same color.

Henry turned to Liza. "What has gotten into you? I mean, you've always been somewhat irate and sarcastic, but ever since New York, you've alternated between lightly cynical humor and downright vindictive abusiveness towards me and towards Malka. I've had it. What the hell is wrong with you?"

"I could ask you the same question," the felinoid retorted. "Ever since Malka rescued you, you've been way more sure of yourself. Stop it. It really doesn't suit you."

"Liza, you have no idea how I act around people I've warmed up to. Honestly, I'm just finding out myself. Besides...," the camp-raised boy turned to Malka, "you ... collected me ... shortly after my parents and I had left our camp. There, I was the only one with any interests besides mining. Everyone treated me with suspicion and contempt. I was insulted, demeaned, and beaten for it. In situations like that, you question whether...."

114

"Whether there really is something wrong with you. If it is not somehow your fault," Malka finished for him. A knowing stare crossed her features.

"Exactly. That doesn't really do wonders for your self-confidence. Like when I told you I could speak other languages? It never occurred to me before that it could be a useful skill. Everyone around me insisted that it was worthless. I began to think that, too. It took me until just last week to think it might be useful to you. Even then, I thought of it only as a last-minute attempt to stay with you. I'm finally starting to see what it's like to have some sense of belonging. If that makes me more confident, then so what?"

He paused, turning back to Liza.

"And don't try to change the subject by making this about me. This is about your behavior. I may be a bit more sure of myself, but here's a little hint: throwing stuff at the people you're supposed to protect? It ain't normal. What's bothering you?"

"What's bothering me? I'll tell you: I'm stuck on a boat with a charge who refuses to take seriously the magnitude of the threat she is facing, who won't adequately prepare for what may be ahead. That's what."

To Henry, it was becoming clear: Liza was convinced of what she said and was actively trying to be hurtful. As the felinoid pronounced the words, the Thag's mouth dropped open.

"After all I have done and all I have sacrificed, how can you say that?" Malka replied. She sounded as if almost in tears. The camp-raised boy interrupted the exchange.

"She's right, Liza. We have a general plan of action, a sound one. But, I have heard you over the past week. You have been personal and scurrilous. That's not very constructive. You can't expect her to predict the future and plan for its every possible twist. It's not reasonable."

"Don't presume to tell me what's reasonable. I know what the Urumi are capable of. I've experienced it. Suffered from it my entire life. As far as I'm concerned, there's no amount of preparation sufficient when it comes to dealing with them. And if that means bashing in a few pieces of furniture to snap some sense into your little friend's head, so be it."

"Okay, then. What exactly did they do to you that gives you the right to judge for us?" he paused, staring pointedly at the felinoid. "Who are you, Liza, really? You've avoided that simple question every time we've asked you. I'm sick of it."

In response, Liza harrumphed and allowed her upper body to lazily flop onto the divan, so that her head dangled over its far edge, out of sight. Her arms dropped out haphazardly to her sides. It was as if she hoped the blasé gesture of dismissal would cause Henry to give up.

"Liza…," the boy pressed.

"Henry, stop prying into matters that don't concern you. It tends to have the general effect of ticking people off. Besides, you don't want to know. Believe me."

"Liza." It was Malka who spoke this time. "I believe that Henry is correct. If what's causing you to act in this manner is due to fear, regarding a specific threat that the Urumi present, we cannot address it unless you tell us what it is."

"What happened to you, Liza?" Henry demanded.

Suddenly, Liza sat up, curling her legs under her as she did so. She rested her hands palms up in her lap. In the space of less than a second, she had gone from the picture of juvenile slovenliness to the figure of a demented Buddha.

"Do you really want to find out?" the felinoid snapped. It was a challenge – a dare – rather than an actual question.

At this, Malka cracked a sad smile. Those were the exact words she had said to Stanley, whenever he had pressed for more information about her own past. They'd had their desired effect. Usually, he would stare at her questioningly for a few seconds. Then, he would change the subject. Malka couldn't really blame Liza for her reluctance to share personal information with them. The Thag was very particular about who she chose to discuss her past with as well. She had been about to demur. Henry, however, beat her to it with a different response:

"*Yes, Goddamnit!*"

Liza sighed. Her mien grew much more serious, melancholic. Clearly, she did not enjoy thinking about what she was about to tell them. She looked down at her hands for a moment, as if preparing herself. Henry and Malka waited, allowing her to take the time she needed. The green-eyed girl looked up and began:

"I have told you that, in your years, I am almost a century old. But, I am at a similar stage in life to your own. That means I've lived, seen, and suffered a lot more than you. I've also implied my species served as the origin for stories about the familiars of those who practice mystical arts."

"Yeah," Henry prompted.

"Except, they're much more than stories. Before the invention of the firearm, my people lived mostly on the western side of the Ural Mountains. Mostly, we kept to ourselves. Occasionally, some of us would partner to work with humans. A number of them practiced arcane arts. Of course, like with any people, there were some who were violent or sought to do harm. But, most of us used our abilities of speed and strength to combat the Urumi. A lot of us used to live among human society, too. Our true natures were widely known. It was no problem. Hell, during the fifteenth century, my parents...."

The felinoid let out a sad chuckle, shaking her head.

"My parents were minor nobility – Boyar and Boyarin – present at the Walachian Court in Târgoviște."

"Wait. Hold on," Henry said with incredulous fascination. "Your parents were members of Vlad Dracula's court?"

"I read a book about him in the Thag's camp. Is it true that he was a vampire?" Malka asked, as if this were a perfectly reasonable question. At this, Liza actually laughed, regaining a small amount of her usual annoyed mirth.

"Oh, please. You humans can have such overactive imaginations." Liza offered a dismissive wave of her hand. "There's no such thing as vampires." Then, the felinoid paused, looking morbidly interested as if just having thought of a question of her own: "What kind of stuff did you read in that camp, anyway?"

Malka listed a dozen or so titles that she had enjoyed. Henry didn't recognize all of them, but he summarized: "It's mostly a mix of rather contemporary Gothic and Romantic fiction." At this, Liza brought her forehead to rest in her left palm.

Seeing this, the brown-haired youth pointed out: "You're acting like were-cats, devil-possessed spirits, mystically powerful gemstones.... Yep! Those are all totally real. But, vampires? Pfff ... don't be crazy.

"You do have to admit, Liza, it was kind of a valid question at this point. But, what happened? How did you end up working for the Society?" Henry asked the question, attempting to keep the conversation on topic. It appeared that Liza would take any opportunity to try and sidetrack the discussion.

"Whatever," Liza replied. Taking another breath, she continued.

"Perhaps, it was because of our speed and strength that the Urumi decided all of us were a threat. They began ordering those they struck deals with to tell no one of their interactions. Instead, they were commanded to blame us for calamitous instances caused by the servants of the Dark Prince."

"Thus, the legends about evil deeds done by witches and their familiars?" Henry interpreted.

"Exactly. These legends became widespread. The situation of my parents became much more uncertain. At first, they simply kept to themselves. As I'm sure you've noticed, we're much faster than any human – except, weirdly, Malka. Any weapon they could bring against us was useless. Instead, it was the humans who got branded as witches. It was they who suffered.

"All that changed after the use of guns became widespread on the continent. Just like that, our advantage disappeared. Many of us heard stories from descendants of the slaves, who'd been brought to our region from India during the Mongol invasions. At least when they had been forced to leave their land, humans there were much more accepting of our kind. Though, the ... breed of my people that lived there was larger, stronger.

"Eventually, with their lives in jeopardy, my parents were left with little choice but to flee. They joined the exodus of the many like them. From all walks of life, their situation in Europe was no longer tenable. They went by foot, first across the steppes of Cossacks' lands, then down into the Caucasus, across Persia...."

"Wait. Hold on again," Henry interjected. "If this actually happened, how come nobody noticed?"

Liza nodded sadly, but knowingly. She responded with a question of her own.

"When you see a bunch of stray cats lying or walking around a field or a town, do you question why they are there, or where they are going?"

"Um, not really, I guess. But why couldn't they have just retained human form and stayed. Or, left on a ship or something."

Again Liza nodded, snorting.

"I'll let you two in on a little secret," she said at length. "I'll take it neither of you two has realized that you've never seen me sleeping in human form. Would you care to hazard a guess as to why?"

"Wait. You mean, you can't?"

"Right. That was the problem. Try to hide your identity around humans and it's only a matter of time until someone sees you change form. Then you're branded as practicing black magic and bang!" The felinoid smacked her palms together, fingers outstretched, to emphasize her point. "Over time, with fewer and fewer of us there, the stories of our existence became just that. Stories. The few of my kind who remain in Europe now are reduced to living as stray cats. That's all they can be with any measure of safety."

"So, any time I see a cat, it's actually one of you?"

"Nope. Sometimes, Henry, a cat is just a cat. Most of the time, in fact."

There was a pause as the two human youths digested what they were being told. Eventually, Malka broke the silence.

"What happened when your parents got to India?" the Thag asked. "When I first found out who you were, I was wary. I'd been told all of my life that your kind was a threat, just as much as the Urumi. That if one of you appeared, it was the sign of a curse."

Again, the same sad nod.

"As you know, a gun can work just about as well against an Urumi as it can against us. Eventually, my parents made it to India. But by then the Urumi were there too, their efficacy in Europe reduced. They did the same thing in India as they had in the place my kind had left, putting the lives of our kind that were native to the subcontinent under increased threat, too. They proved unwilling to offer assistance as my parents and those like them arrived. In fact, they turned against us. We couldn't fight them, even though we're more agile. They're bigger, stronger, like a panther or a tiger. We were the newcomers.

"Like most, my parents ended up in a camp. In their case, one carved out deep in the forest of Uttar Pradesh. It was constantly assailed both by the Urumi and those of our own species that lived there. That's where I was born. Some of my earliest memories are of growing up in its mud and squalor. There was never enough to eat; it was even more dangerous to leave than it was inside. My parents were killed when I was about eight, in terms of human development. The Urumi ambushed them while they foraged outside of camp. I was left alone with my younger brother."

The felinoid laughed softly at her recollections.

"He was always so positive. I don't know how he did it. We were constantly under threat of attack. With resources so tight, it was all we could do to fend off assaults from others within our own camp. You'd have thought we'd have come together as one, united in our misery. But, our camp was so fractured according to the places we'd come from in Europe. We could barely mount a coherent defense. That was the life from which the Society plucked me and my brother over ten years ago. Now, I fear that camp is gone."

Both Henry and Malka sat, spellbound.

"So," Liza finished, fixing them both with a cold stare, "when I say there's no way that the threat posed by the Urumi can be taken seriously enough, I'm uniquely qualified to make that determination. Ha! Without the Urumi, I might've been born heir to a noble title. As it was, I grew up in the midst of strife and itinerant poverty, displaced from a home I'd never even visited. They did not only rob me and my parents of that home, they attempted to destroy my entire species by setting us upon each other."

Silence filled the room for at least a minute after Liza finished. What she had told them seemed too farfetched to believe. But, seeing the things that they had, Malka and Henry found that they were in little position to deny it. It was the Thag who spoke first.

"Liza, I understand why this would make you fear and hate the Urumi to such a great extent. But, it still does not make sense. You have always been keen on planning strategy against them, but why have you been so derisive to us recently? Is it because we must return to Europe? Is it safe for you there?"

"No. And no less so than in America," the felinoid responded. Looking across the room at her charge, she continued. "I've been back to Europe a few times during my ... tenure with the Society. During those, I've found – ironically – that because very few people believe that my kind even exists anymore, if someone were to see me change form, they wouldn't believe their eyes. Or they wouldn't be believed by others. Still, I wouldn't exactly be able to live a full urban life there."

"Fine. Then what?" It was Henry asking the question this time. The blue-eyed boy was beginning to get the sneaking suspicion that the felinoid was going on at length about her species' history, in order to avoid discussion of what was truly bothering her. Malka was right. Liza's personal family history might explain the intensity of her hate for the Shadow Warriors. It didn't explain her recent vitriolic attitude.

"It's nothing."

"People or, uh, cat ... things...," Henry replied, scarcely believing the words that were coming out of his mouth, "Don't throw clocks into walls because of nothing."

"Felinoids," Liza corrected. "We used to be called felinoids. And for all you know, maybe we do." The black-haired girl looked at Henry with an enigmatic smirk and raised her eyebrows twice, quickly.

"Liza...," Henry groaned in a tone indicating he was not convinced.

"All right. Fine." Liza stopped for a moment before continuing. She addressed her initial comments to Malka. "I'm not sure how much you know about the Society's role in helping rescue Nell in India?"

Henry scowled. Another non sequitur.

"Nell told me what happened shortly after the Invisible Circus, where my sister...," the Thag trailed off.

Liza nodded. She pressed on, sounding as if determined to get this over with.

"You know that Nell was rescued from one of your Thags by a large black pantherlike…."

"Balu," the Thag confirmed, recalling the name. "He was with another boy. Dark-skinned…."

"Yeah. Tiku," the black-haired girl cut her charge off.

"You know him?" the Thag asked.

"Um, yeah. He's my kid brother," Liza said it as if this were something immediately obvious.

"Oh."

"Uh, wait. If he's your brother, why difference in…," Henry asked, unable to control his curiosity.

"Skin tone?" The milk-skinned felinoid anticipated the question. "You know when a cat gives birth to kittens and they all have different fur patterns? Works the same with us. In his feline form, Tiku's a thick-coated, orange-striped tabby. There's no rhyme or reason to how we look and whether we're related."

"Okay. So what does all this have to do with…." Henry suspected that the felinoid had again managed to sidetrack the discussion. It was turning out that getting Liza to address her feelings was proving to be about as easy as, well, herding cats.

"I'm getting to that," Liza snapped, annoyed she had again ultimately failed to drag the conversation off track.

"The Society mostly uses my kind for recon and asset protection," she explained. "The pairing you noticed with Tiku and Balu is a standard operational structure. My breed is fast, small, and stealthy. The larger natives of India and Africa make up for what we lack in brute power. It makes sense. But in light of the strife between our two races, I think Arunesh and Zitar – who functionally are the heads of the Society – have continued to favor it as a means for fostering understanding between…."

"Liza? Does this have anything to do…," the camp-born boy interrupted, becoming increasingly impatient with the large amounts of irrelevant information.

"Take a look around you, Henry," the green-eyed girl challenged, as if the youth was being deliberately obtuse. Obligingly, the boy looked around the room. "What's missing?" she asked.

"You mean aside from a clock case and an upright coffee table?" Henry couldn't resist. As he spoke, he realized that some of the felinoid's sarcastic wit must have been rubbing off on him.

"Haven't you noticed what I've told you? My brother worked with a partner. I'm here by myself. Why'd you suppose that might be?"

"I don't know. Because black Indian panthers would stick out like a sore thumb in the American West?" Henry was unable to keep an acerbic tone out of his voice. Honestly, he still couldn't see what relevance the Society's operational organization decisions had to do with what he had been trying to get the felinoid to discuss.

"Henry." It was Malka. She put forth a hand, indicating that he should back off. Then the Thag interpreted. "You had a partner, didn't you? Something happened."

Liza sighed deeply, recalling a memory that was more painful than most. Finally, she addressed the issue directly:

"My partner, Tem-vi, and I worked together for over ten years. The Society placed us together from opposing sides of our conflict. When first recruited, we were sent out on missions without any training or time to get used to our new roles. It was their policy, they told us. For a while, we were unable to overcome our mutual mistrust; our first few assignments ended in blunders. Yet, that was enough. It gave us a clear purpose. We pulled together. Slowly, after being assigned to more missions, facing so many crises together, closeness developed between us. It was a kind of brotherhood. We never questioned that we'd guard each other with our lives. Through so much itinerancy and struggle, which became all the certainty we had, it was all we needed. It was as if we had our own world that we could shape to fit our own lives.

"But, as always seems to happen, disaster strikes when you least expect. It happened just over two years ago. Though, it seems like far less."

Liza paused at length, taking in a deep breath, then blowing it out again.

"What happened?" Malka asked, riveted on Liza's recounting. The felinoid breathed in again.

"We were on a simple recon assignment: observe a pair of humans, who had apparently gotten lost in the wild. Report back. We didn't have any other information. But, as I'm sure you've figured out by now, the Society tends to be less than forthcoming with that kind of stuff. So, that was all we knew.

"They were young. One – a boy – was a bit younger than we are now. I'd guess the other seemed barely under ten. They were traveling with a trio of other youths, but the two we were supposed to be watching rarely left each other's sides. They were camped somewhat near a verdant canyon in the wilderness, where we had ensconced ourselves. Our subjects had moved to the canyon's edge, in order to follow a certain species of bird. Apparently, they found its song pleasing. We could barely see them as they moved. Tem-vi wanted to advance further, to get a better view. I told him it was too risky. Our subjects might see us if they turned back." Liza chuckled softly. "But, did he listen? No. Of course not. I let it go. Figured that if he ran into them, the worst that could happen was he'd give them a scare.

"Tem-vi, in his pantherlike form, had made it about halfway up the path they'd taken, when he encountered the girl. She was walking back on the path to their camp by herself. There was a bunch of flowers in her hand. She dropped them. At the time, I remember thinking how unusual it was: the older boy rarely let her go off on her own. Tem-vi and the girl just stood there, staring at each other. I'm not sure who was more surprised."

The green-eyed girl paused again, taking in a ragged breath. Her eyes began to water.

"What happened to him?" pressed Malka.

"Did your subject – this girl – somehow hurt Tem-vi?" Henry inquired, with a bit of doubt in his tone. The felinoid shook her head, looking downward. Then she looked up again, pressing forward.

"Then, the other one, the older boy, came down the path after her. We'd just started observing them the day before. Neither of us noticed it. If I'd been paying closer attention, maybe I would have: he'd had a gun – a rifle – in his belongings. He saw the two of them staring at each other and...."

Liza paused, her increasingly shaky voice cracking. "He decided to shoot first and ask questions later. He shot Tem-vi." The felinoid let out a ragged breath.

"And that wasn't the end of it. Having heard the shot, those they had been traveling with – locals – came, running. The boy who'd killed my partner handed one of them a knife. I don't speak the local language, so I could not tell what was said. But it looked as if my subject wanted him to flay Tem-vi's corpse. They'd brought a horse to the site. His body was dragged away unceremoniously behind it. I didn't follow. I didn't want to see any more. That was the last I saw of him."

Another moment of silence. Tentatively, Henry took voice.

"Liza, I'm sorry. That must have been extremely difficult to go through. Even after everything else that you've seen. But I have to ask: why are you mad at us over it?"

The green-eyed girl looked up at the boy.

"Don't you get it?" she hissed. "After that, I insisted on working alone. I went from assignment to assignment. Didn't take any time off at all. As long as I was focused on the task directly in front of my nose, I didn't have to think about what had happened. I didn't have any attachments. That was the way I liked it. It was just ... less painful.

"But since then, none of my missions has gone on this long. After we retrieved you in New York, I began to realize I was growing accustomed to the company of you two. That was the last thing I wanted to happen, thank you very much. You – or at least Malka – were assets to be protected. Nothing more. I didn't want to see you any other way. Not only because I have seen what the Urumi are capable of, but also because I was worried that I wouldn't be able to protect you. I didn't want to like you. I didn't want to get close. Yet, even as that happened, I convinced myself that I didn't want you to like me. After what happened, the thought of belonging again scared me."

Liza let out a short bark.

"You know? Sometimes I envy you two. Having never had ties to anyone? You should consider yourselves lucky. That way, you can never lose."

A stunned silence fell over the stateroom again. Malka eventually followed with a perspective of her own. For the first time, she felt as if she were speaking in the capacity of Master:

"For all of my life, in my own camp, the thing I wished for most was to be one of them," Malka finally told them. "When I met Stanley, I envied him. At least he had one other person with whom he could identify. But as much as I wanted those things, I could not force them to be so. I envy *you*, Liza, in an odd way. You at least had two caring parents, a brother and someone with whom you could share a true understanding, if only for a time. From my life, I can tell you that feeling completely alone – all of the time – eats at you. My memories of it still do. I would not wish that life on you.

"I also do not think that you can control with whom you identify. We have spent much time together and shared a common goal. Now that I think about it, we share similar backgrounds. Yet, I did not much care for you, when I first made your acquaintance."

"Yeah. Same here. Goes for both of you, actually," Henry chimed in. Malka continued, smoothly.

"However, I have grown accustomed to you. It is natural. As, sadly, was my isolation in the camp. My inclusion was not something that could have been forced. It was not something I possessed the power to choose. In different ways, we both suffer from that reality of – for lack of a better term – human nature. I cannot say whose purgatory is better."

Henry shrugged, allowing the Thag to finish her contribution.

"Basically, what Malka said," he added.

Liza smirked, a frown on her face. At length she said:

"That's it? Just 'that's life'? I guess.... Oh, hell. If there's no way around it, why does it have to *suck* so much?"

Both Henry and Malka stifled a laugh at the felinoid's frank analysis of the situation. Then they turned serious again.

The Thag sat for a moment. She thought of Stanley. How he had spoken fondly of his adventures in Africa. How, then, his possibilities had seemed limitless, despite the peril in which he had found himself. Despite her own tightlipped-ness, Malka had come to feel a sort of camaraderie with Stanley and Mungo in India. Then, that had ended horribly. And now, here she was, half a world away with two people who she had met, seemingly by chance. The stakes of their common quest were high. The outcome uncertain. And yet, as Liza had mentioned, it was as if those circumstances created their own space. Where they could dare to dream of the impossible and achieve it. Where the consequences of failure were both acutely felt – and yet – were not countenanced. Where they possessed the fleeting power to remold world in their image.

Eventually Malka answered, speaking to both in the room.

"These times when commonality is forged, they are rare. Often, they come in the worst of circumstances, yet they lend an audacious hope. That hope is scarcely noticed when present. It can only be mourned once gone. Nothing can be served by fear of what lies ahead. Through the lens of hindsight, what has transpired now will, no doubt, seem better than whatever is to come."

Another silence. Malka's words had not been meant to offer a solution. They were merely a recognition of the situation.

Finally, Henry spoke.

"Well, that's not depressing at all." A short pause. Then: "Right. So ... dinner?"

"After all this, that's all you've got to say? Let's have dinner?" Liza asked, half in disbelief.

"Yeah. Why not? I'm hungry." As one, the other two present in the room shrugged.

"Sure," they both said.

Liza got up from the divan and called the porter to place their orders. Then, all three of them moved out to the parlor's balcony to watch the sunset. Until long after night fell, they chatted easily about whatever came to their minds. Nothing – well, almost nothing – was taboo. The future was uncertain. The stakes high. The odds were against them. But, they allowed themselves, at least for now, to be overcome by the precariously unwavering certainty that whatever trials the future presented, they would be able to surmount them together. What if they failed? Until that happened, fear of reality had no place among them.

Fourteen

The grotto was silent. That was the way Bozhena preferred it. The blue-eyed Slav waited for her Chosen to arrive. The warrioress did not wish to be here. Yet, she couldn't put off this meeting any longer. Eventually, she would have to report to Ziya. It was best to get it over with.

The Urumi remained perplexed by what had transpired. It should not have been possible. Before her appointment with the Fragment's holder, Bozhena had not wished to see the young boy she had kidnapped dead or consigned to serve the Dark Prince. Nor had she wished that fate upon the one who guarded the Fragment. It had not mattered, of course. Being a victim of the Transmutation herself, Bozhena's will was no longer her own. She'd made no attempt to control her emotions.

She had tried to make the best of an otherwise dismal situation when constructing her plan for victory. Once, her mother had taken Bozhena and her sister across the border to the great Skarbkowski opera house in the Hapsburg-controlled city of Lwów. The older woman had wanted to show them Stanisław Moniuszko's Straszny Dwór, an opera that had been banned in imperial Russia by the Czar's censor. The girl had loved it, and its message. That had been two years before her own consignment to the service of the Urumi. During that time, she'd gone to the theater whenever she'd had the chance. Even now, whenever she could, Bozhena used her abilities of instantaneous movement to attend opera performances in cities from Budapest to Paris.

She'd decided to use the Invisible Circus to have a bit of fun, when designing the trap she'd set for the Thag. While the Urumi controlled the Invisible Circus, they did not produce content for it. Instead, the idea was to use the work or arts of others to discover information that they could use. Having discretion over what to show, the blond-haired girl had decided to use its mystical forum to display an opera that she had not yet seen – and a production of it that would not take place for years. Partially, she had done this for her own enjoyment. But, she had not chosen the work at random. The Urumi had intended to use the plot's connection to her target's past to unsettle her and make her question her own resolve. The idea was to make the girl more susceptible to the demands of Bozhena's dark figure.

They were demands that she had not wanted to make. Yet, she felt the overwhelming urge within: comply with the Dark Prince's wishes. She could not do otherwise. It made her feel miserable – alone. As the production started, Bozhena had found that she was crying.

Then, it had happened. The warrioress had been planning to wait until a later point in the show. But, with the music building, the blond-haired girl had actually acted on an impulse. This went against her plan. As such, it did not serve the Dark Prince by increasing her chances of gaining the Fragment. In fact, quite the opposite. Even more surprising, instead of arguing and following through with her threats, Bozhena had managed to only offer token demands. Every fiber of her instinct screamed for her to serve the Dark Prince. She had been unable to resist the urge to fight when her demands were refused.

Then the music resolved into a melodious line: a duet. Knowing the story, Bozhena could not get out of her head how similar the situation being played out onstage was to the one that existed between her and her target. Being aware of the subcontinental native's origins, she knew: had they been allowed to discuss freely, they likely would have found much in common. Instead, their first encounter in that small Madras temple had destined them both for a more adversarial relationship. With every ounce of her being, which she had thought was no longer hers, Bozhena wished that it did not have to be so.

The impossible happened. As a break in the melody occurred, Bozhena realized that she stood before her target in her human form. She had no idea how it happened, either. Urumi showed their true forms only to each other. They were constitutionally incapable of doing otherwise. Yet, Bozhena did. Voices in her mind screamed with confusion over what had happened. They also chafed with the certain knowledge that she did not wish to fight the individual who stood before her. Nor did she want to see the destruction that would be wrought by the Urumi if the Fragment fell into her possession. The

Slav had no desire to serve the Dark Prince, nor anyone else. She was not supposed to have any choice in the matter.

She'd intended on using the boy against her target. And yet – somehow – she had managed to tell the dark-skinned girl to take the youth. In its turmoil, her mind overcame the imperatives that had been emplaced by the Transmutation. If she had still needed to eat, the strain would have caused Bozhena to vomit.

Even more strange, those imperatives demanded that she confer immediately with her Chosen and accept whatever punishment he may decree. Bozhena found that she had been able to not comply. Instead, she'd acted on a whim; since the mystical opera, the blue-eyed Slav had spent most of the time around the family manor where she had grown up. It now sat empty, cold. It was as much a shadow of its former glory as Bozhena felt a shadow of her former self.

There, Bozhena found that her impossible freedom became easier. She grew more accustomed to its presence. At first, her actions had been taken out of the sheer power of emotional strain. After a couple of weeks, the blond warrioress found that she was able to defy directives through mere conscious decision.

None of that changed the fact that she remained an Urumi, obliged to serve the Dark Prince. The blue-eyed girl could not simply quit. The longer she waited, the more likely it was that Ziya would come looking for her, demanding an update on her progress. Eventually, he would find her – not even attempting to fulfill her mission. If he suspected that something was amiss, it was likely that he would have her killed.

And so, she had elected to bite the bullet. The Slav knew that she could now dissemble regarding what had gone on with her mission – itself an act of free will. But after meeting with whatever punishment Ziya's ire dictated, she would be assigned more tasks. To keep up her guise of compliance, she would have to follow some of them. Further, knowing Ziya, those tasks were likely to prove ineffective and distasteful. She resented her position within the Urumi's ranks. If she had to be consigned to their service, why could she not receive a position in keeping with her clearly greater ability? She chafed against the Chosen and his Prince. But, Bozhena knew that in the longer run, she would have to continue serving them. She could only allow what had happened at the Invisible Circus to be a momentary lapse, if she was to survive.

Bozhena felt the air move behind her. She turned from the oriental altarpiece that she had been regarding. Ziya al-Din entered the grotto. His dark features scowled. The young commander walked up to the fair-skinned Slav, backhanding her across the face without warning.

"Where have you been?" he thundered in his oddly high voice. "We have been searching for you."

Bozhena kept her face pointed downward, in a gesture of outward submissiveness.

"My Chosen, I have been about the mission that you, in your wisdom, have assigned me."

"Very well. Where is the Fragment? Our new recruits?"

"I regret to inform you, my Chosen. My plan has failed. The one who guards the Fragment is gifted with abilities of speed and strength that rival our own. She has again bested me in combat. The Fragment remains with her."

"Unacceptable!" he roared. "No human should be able to match us in combat. You simply lack the proper attitude. You do as commanded only because the Transmutation demands that you do so. You have no real wish to serve us. You do not see it for the true gift that it is. It disgusts me." He punched her in the gut. Bozhena allowed herself to double over, falling to the stone floor. The blow had not affected her that strongly. However, she allowed it to appear so.

"You are fortunate I do not order you killed where you lie!" he raged, turning from her. The Sudanese Arab stood, fuming for a minute. Then, he asked:

"I will at least assume that you did not report back promptly because you needed to follow your targets' continued movements, in order to set a new trap?"

Bozhena hesitated a moment.

"No, Chosen. I was too injured afterwards to pursue. I have been recuperating since that time."

Ziya turned back to her. Bozhena remained in a pile on the floor.

"This is an outrage. Even fear of my punishments proves insufficient to motivate you! I entrust you with one of the most important missions commanded unto us by our Dark Prince. It has been a blunder because of your incompetence."

Bozhena remained silent. She knew that any successes, which this impossibly difficult mission had encountered, had been *because* of her. Ziya was simply looking for a convenient outlet at which to vent his frustrations. Seeing her lack of response, the Urumi's commander continued. There was increasing fury in his voice as he spoke.

"You have taken quite long to produce no results. You have allowed two of my most trusted lieutenants to be slaughtered. You have failed in your ploy, which you promised would gain the Fragment and provide us with two new recruits. All of this time, you have done nothing but question my sacred commands. Now I find that you have again lost your targets' trail? Very well."

Ziya began to strut back and forth as he explained. His tone suggested that he was exceedingly pleased with himself.

"I have selected a new lieutenant, one who will help achieve our goals. The Dark Prince calls to me. He grows impatient. We shall do as He commands us, in His glory, together."

Bozhena looked up, a hopeful, questioning expression on her features. Could it be true? Had Ziya been deriding her performance merely for his own amusement? Had his intention all along been to promote her?

The Chosen noticed this as he strutted back towards her. The eldest son of the Mahdi laughed harshly.

"Did you really think that I meant you? How pathetic! On the contrary. I have selected one whose zeal for violence is exceptional and whose appetite for brutality – like my own – is truly insatiable. It was the strength of her angst – against her own people, no less – that called us to her. She is one of the very few to have entered our service willingly. Commendable, isn't it? Try as I may, I cannot imagine you doing the same.

"From the time of her Transmutation, she has proven far more useful than you. While you preoccupied your attention with your father's activities in Pondicherry, it was my new aide who discovered that the one who now holds the Fragment had attempted to hand it off to her sister. You wasted time attacking the Thag's red-haired friend. My new lieutenant delivered your current target's sibling to your father. It was due to this new one's spite that my previous plan to unleash the object's power nearly succeeded. You hindered it. While you were hiding in the burnt-out ruins of some castle, it was she who activated our asset in the British Isles. She performed an exemplary job there as well. It was your mission, but I am afraid that you were not anywhere to be found when a task required doing."

Again, Bozhena was incensed. If she had killed the son of the now-deceased police chief, along with his friends, they would not have been able to thwart the Invisible Circus. As for missing the hand-off of the Fragment? Not even the Urumi could be in two places at once. It was a seeming impossibility that the Thag could have produced an identical replica apparently from nowhere. What reason would she have had to suspect? Eventually, the trade-off would have been found out. The Chosen's plan – even though she had hated carrying it out – would have moved forward successfully. As it was, she only responded:

"I understand, My Chosen. I am humbled by your worlds. I shall return to New York and attempt, once again, to discern my targets' trail. They were staying at an upscale hotel in New York City. I shall make inquiries of the staff there...."

Ziya cut her off.

"You will do no such thing. I can see that this mission is far too important for us to entrust it to one so incapable as yourself. My new lieutenant and I will see to the matter personally. You will return to shadow the other youth who thwarted our attempts to unleash the Fragment's power at the Invisible Circus. From your previous reconnaissance of him, you know his current location."

"Apologies, my Chosen, but I remain practically certain that he does not – or will not – play any further role in our plans, whatever they may be. He remains nowhere near the object we seek."

"This is true, for now. However, I possess knowledge of things, which you do not," Ziya responded. "As is the case with our British asset, it remains prudent to keep measures in place to thwart that youth. He has worked against us in the past. For that reason alone, he should be kept under surveillance as long as we have the resources. You may serve in that capacity."

Bozhena could not believe what she was hearing. Not only had she been passed over for promotion, but she had also been removed from her mission and commanded to shadow one who most likely no longer played a role in any of the Urumi's plans. The Chosen – whose position she envied – no longer regarded her as important enough to order her beaten or killed. Ziya had simply declared her irrelevant.

In that moment, Bozhena made a decision. She had been assigned to shadow one youth near the Alps as a make-work assignment. The Slav would see to it he became very relevant, indeed, to the Dark Prince's plans. The blue-eyed girl knew that she still possessed the means to do so. But, she would not take these actions to aid the Dark Prince, the Urumi, or their Chosen. She hated them all. They had used her father's twisted ambition to rob her of a life. Then, they had refused to reward her tireless, if unwilling, service. Not only did she owe them nothing, but she also despised them. Now that she could, the blond-haired girl wanted to punish them for it all. Bozhena might die in the process, but the blue-eyed Slav realized that she no longer cared. Vengeance now mattered more than the shadow of a life that she led.

Of course, the blond-haired girl let none of this show to the Urumi's commander. Instead, she demurely stated: "Of course, my Chosen. It shall be as you have ordered."

Bozhena stood. She began to move her cape about her, preparing to remove herself to Fribourg. However, the Chosen raised his right hand in a gesture, indicating that she should wait.

"There is one final thing: I wish for you to meet with the one who now serves as my lieutenant, so that you will know to whom you are required to listen and obey." Ziya raised his voice. "Enter," he called.

The Chosen's new lieutenant entered with a flourish, needlessly making loud footfalls that echoed off the stone floor. She moved to regard the Slav with that look of superior disdain reserved for those deemed inherently below one's rightful position.

Keeping her expression carefully neutral, Bozhena stared back into the brown-eyes of the one who unknowingly now held only the mere power to issue her orders, commands that she was no longer obliged to obey.

The Chosen's new top aide regarded the Slav for a full moment before speaking. It was a simple question meant to reinforce who was in charge:

"I take it that you finally understand your place now … Bozhena, isn't it?"

The blue-eyed Slav nodded. She remained standing. Her expression betrayed nothing.

"Good. If I were you, I'd best be on my way."

In a blink, Bozhena was gone.

Her dark skin only a few shades lighter than the black trappings of the Urumi that she now wore, Zaima cackled at the seeming compliance of her newest subordinate.

Fifteen

They waited behind a bush. The dampness was almost suffocating. Malka had grown up in a humid environment. But, this was different. The dimness brought by the tree leaves that dominated the outdoor enclosure, combined with their prevention of any ventilation, made the space unwelcoming.

Shortly after their arrival at Fribourg's main train station, Malka, Henry, and Liza had made their way to St. Nicholas Cathedral. Next to it, they had immediately discerned the buildings that comprised the institution's Catholic school. The three had remained in one group, not wanting to split up, should one of them encounter a member of the Shadow Warriors. Scouting around the cobblestone streets, the trio quickly discovered the school's dormitory. Henry had been the first to see the sign, it being only in French. As he had promised, the boy's Francophone abilities had proved much the boon since their arrival in Europe.

They had disembarked the Rhynland in Antwerp. As first-class passengers, they were exempt from passport control and had moved easily to hire a carriage that would take them to the city's train station. There had been a bit of confusion at first. It appeared to both Liza and her charge that most of the carriage drivers seemed to not understand Henry's French. They had become slightly suspicious, wondering if his language abilities were what he'd claimed. The brown-haired boy explained: in this part of Belgium, most people preferred to speak "like the Dutch." At least, that was how the camp-raised boy had described it.

Initial difficulties aside, they and their belongings were finally loaded onto one of the carriages. It took them from the Red Star Line's quay to the city's main railway station. On the way, they'd passed through Antwerp's city center. There, Malka and Henry had gotten their first taste of European architecture. They looked at the ornate buildings, spellbound. The edifices were like nothing Malka had ever imagined. For Henry's part, he had read of such buildings, but he had never dreamed that he would see them in person.

"This," he'd said as they passed, "is culture."

Liza rolled her eyes.

Eventually, they'd arrived at the terminal of the Antwerp-Mechelen-Brussels railway line. Inside the wooden station's reception hall, Malka had committed a minor blunder. When changing money, she had, at first, flopped down an entire saddlebag of cash.

"Don't change that much!" the felinoid hissed.

"Why not?" Malka asked, confused.

In response, the black-haired girl had drawn her lips close to her charge's ear.

"Because, for one, I doubt – strongly – that you have any idea of how much that can buy. Also, we just want enough currency to get us an appropriately nice train ticket and pay the carriage driver."

"But we will need more after. Why not change it now?"

"Because, after we leave Belgium, nobody's going to want Belgian francs." At this, the Thag's brow furrowed.

"How many different kinds of paper will we need to get where we are going?"

"Just to where your friend is? Three, at least. Then, assuming we're going somewhere in Central Europe? At least another, maybe more."

"But, why, Liza?" the Thag asked.

"Because the people here speak many different languages. Each group has its own country. Well, kind of. And they fight. Constantly."

The blue-eyed girl remained perplexed.

"In India, people spoke many different languages as well. The paper was the same."

The green-eyed young woman snorted, shaking her head.

"Yeah, well. Out of the mouths of babes and all that."

"You have told me that, in terms of our development, you are no older than I."

"Right. That's not the point. Welcome to Europe. It's a whole different game."

As it turned out, Malka changed only a small ream of the American cash. With it, they had been able to purchase tickets to Fribourg in a private car.

A bit of a minor surprise had come when the train stopped after only a few hours. They'd been informed that they must disembark. Their tickets allowed them to carry onwards, but that was not the problem. Apparently, Henry had interpreted, there was no central train station in the Belgian capital. They would have to make their way to one on the opposite side of town in less than an hour, if they were to make the service on which they had been booked.

Quickly they had flagged down another carriage. This time it was open-air, rather than having a covered compartment. Henry chatted excitedly with the driver. The man was dressed in a suit, with longish hair and a manicured mustache. At one point, the blue-eyed boy turned to his traveling companions.

"Apparently, this country is made up of two different peoples," he explained. "They speak different languages. Don't particularly like each other, either. But somehow, they make it work. It's kind of live and let live, actually. The problem is that when they have to deal with each other, everything becomes a ridiculously complicated compromise. Uh, did you know the train station we're going to has two names? Like everything in this city? The two language groupings can't even agree on a common name."

"Sounds confusing," drawled Liza.

"It is. In more ways than that," Henry replied. "It's starting to make my head hurt, actually." He paused, cocking his head backwards at their driver. "But that guy? He seems oddly proud of it. Like, in their discord there's some sense of nationhood, even if they can't find it from any other source."

Here, they passed through the main square of Brussels. Both Malka and Henry were suitably impressed. Liza appeared bored.

"In the camp of the Thags, I never imagined such things existed," Malka said as she took in the sight, dressed as she was in a plain white dress and, of course, her sash.

"This is a truly civilized place," said Henry.

Liza scoffed. "Civilized? You know, Henry, what you like to say about being accepted for who you are? Try explaining it to this country. Let alone to the rest of the damn continent."

Depending on which local language one spoke, the trio made it to Brussels' Middle/Southern train station, just in time to catch their train. It took them as far as Basel, where they had been obliged to change again, for a service that took them to Fribourg. With Henry's help, finding Stas's dormitory had not presented a major problem.

Now, Malka moved slightly. Liza, in her feline form, exited the dorm's main entrance. The black cat slunk across the courtyard. It took refuge behind the same large bush where Malka and Henry hid. In a flash, a milk-white skinned girl crouched before them.

"I thought I'd have to wait forever for that friar to take a pee break," she groused by way of greeting.

"Did you find the room number?"

"Got it. Two-oh-four."

Looking about them to make sure that no one was entering or exiting the courtyard, the trio moved back out onto the cobblestone street. They followed Malka's lead. The group turned right, downhill, then right again into an alleyway. Above them, the side of the dormitory facilities that faced toward the river could be seen. It was dusk, a bit after nine in the evening. Undoing the bloodred sash from her waist, the darker-skinned girl affixed the brass doorknob to it.

Surmising what the camp-raised girl was planning, Henry turned to Liza.

"Um, am I missing something here? Or, wouldn't it just be easier to walk in and ask that friar at the front desk about Stanley? I mean, I get we have to keep a low profile, but if we get caught trying to break into the dorm, isn't that going to mean a lot more unwanted attention?"

The felinoid smirked in response. She turned to him, her face a picture of mock disapproval.

"I'm surprised at you, Henry. You've spent this much time around Malka, and you still haven't figured out how she can sometimes feel about making sense? Also, don't forget she's a classically trained killer and a thief. Maybe breaking and entering is just more in her comfort zone. Besides, I trust her not to get caught. That means no one at all will see us."

"Uh, right."

While this exchange occurred, Malka moved the weighted end of her sash in a circular manner as fast as she could. Eventually, the blue-eyed girl let it go, allowing it to unfurl to its farthest extent. It sailed upward, catching on an iron gutter pipe that extended from the building's roof. She tugged on it, trying the hold. It was firm. Placing her feet on the exterior wall of the structure, she began to ascend.

Eventually, the Thag made it to the relevant window. She looked inside. The room was empty. Henry saw her open the window, although from his particular vantage point he could not see how. The Thag nodded. One by one, the other two in her party followed suit.

Sixteen

Things happened almost too quickly to process. The second Stas entered his dorm room, he found himself pinned down behind a length of fabric. It was pulled tight across his neck. Its presence restrained him for only a few seconds. Then it let up, and was removed.

Stas surveyed the scene. A brown-haired boy, whom he did not know, sat unceremoniously on Jurgen's bed, hands folded in his lap. Jurgen, still dressed in his formal attire, lay on the floor. The pieces of his walking stick lay shattered about the room. Apparently, it had been broken when the Swiss-born youth had returned to the space earlier than Stas had. His form lay unmoving.

Presently, another figure moved in front of him. Stas realized that he recognized her, although he had no idea how it was possible that this person had come to be standing in his Swiss dormitory accommodations, thousands of kilometers from where he had first met her.

"M-Malka?" Stas whispered in disbelief. He could scarcely believe what his eyes told him.

"It is I," the camp-raised girl confirmed her identity.

"Why…. *How* are you here?" the Egypt-born youth whispered, still trying to make sense of the situation.

"I …. we believe that you may be required to assist in the quest on which we now find ourselves. As for how I have come to be here? That would require a long telling. There is little time."

"W-what?" He looked around the room again. His gaze settled on the form of his roommate. Then, on the individual who now sat on his roommate's bed.

"What about... Who is...?"

"Don't worry, I merely strangled the one who entered this space before you into unconsciousness. That," Malka inclined her head towards the youth who was seated on the bed, "is Henry."

The camp-raised boy lifted a hand in a gesture of greeting.

"He travels with us now. Also, you are going to need to meet…."

Stas now noticed that a black cat sat perched on the sill outside the room's open window. It had been keeping watch over the street below. It turned its head. The animal's luminous green irises widened visibly. It hissed, turned, and leapt into the room. What had been a quadruped changed form in midair, landing on her feet in the form of a young, light-skinned woman. Stas stared at the apparition in shock. Nell's protector barely had time to process that this creature was advancing on him, intent on doing harm.

"You!" The single word dripped with the vehemence of accusation as Liza lunged for the half-Slavic youth. She was intending to tackle him.

Suddenly, a length of bloodred fabric found itself slung around Liza's waist. It pulled taut, yanking the felinoid away from the one she'd been intending to assault. She struggled. The sash's lengths would not let her go. The felinoid changed forms again, allowing the smaller size of her feline figure to disentangle herself from the entrapment of Malka's sash. Again, the black creature crouched to leap in Stas's direction.

Just as quickly, Malka leaped down. She scooped the cat up and held it under its forelegs. The rest of its body stretched below. It hissed and meowed stridently. The Thag's and the felinoid's faces were pointed toward each other.

"Liza. That was not very intelligent. I do not know what you think is the matter. Nothing will be served by this. You need to calm down. Now."

"Yeah, really," Henry chimed in.

Eventually, the cat stopped struggling. Malka placed it on Stas's bed. In an instant, as the Egypt-born youth watched, still shocked, the young black-haired girl became apparent once more. This time her head was in her hands.

"Liza! What the…," Henry yelled, his voice making clear his confusion and rapprochement. He was, however, unable to make it to the end of his sentence. Liza looked up suddenly, cutting him off. He noticed that her face was a mask of anguish. It was streaked with tears.

"It's *him!*" she managed, her voice ragged.

"It's who, Liza? That is just Stanley. The one we came all this way to find." The Thag's Master questioned her protector.

"It's him. He shot Tem-vi." Her voice sounded like gravel.

Stas looked confusedly back and forth between the three intruders. Finally, he managed to interrupt. The youth, to whom the world had seemed so simple in Port Said, had no idea what was happening.

"What? Shot whom? What is going on here?"

"Liza." It was Henry. He sat on the bed opposite her. "Are you sure? I mean, what are the odds...."

"Do I look like a goddamn mathematician to you?" the felinoid thundered. "It's him. I'm certain."

"Great. Just perfect," Henry said in response.

"I'm who?" Stas stammered. "What is happening? Who and what are you people?"

Keeping the corner of her eye on the anguished felinoid, Malka turned to regard Stas. Henry remained quiet, apparently deciding to let the subcontinental native, whom Stas already knew, take the lead.

"This is the group I now travel with. Liza...," she indicated the black-haired girl, "works for the Society – like Tiku. Apparently, she believes that you killed the partner with whom she used to work."

"What? When?" In response, Liza hissed at him. She appeared even further incensed that Stas appeared not to recall the incident immediately.

"Yes. It would have happened about two years ago. Do you remember an instance in which Nell was alone for a short while. Then, she encountered a large pantherlike figure?"

Stas nodded. "It happened towards the end of our adventure in Africa. We were traveling through uncharted wilderness. A kind of forest. We had made camp. But we'd wandered away from it. Nell liked observing the wildlife. I needed to go hunting, but Nell insisted that she could walk back to camp by herself. I let her, at first. But then I quickly changed my mind and ran after her.

"I found her staring face-to-face with a large cat. She was clearly in danger. I raised my rifle's sight to my eye and dropped the animal." Given the situation, Stas decided not to mention what had happened after. Wanting to scalp the animal, he had used a horse to haul it back to camp. Upon seeing it, the elephant, King, had thrown its carcass against a tree and then trampled it into a pulp.

"You *killed* him!" Liza yelled again.

"Killed *who*?" Stas yelled, still not comprehending.

"Her partner," Malka interpreted. "They had worked together for years. It appears as if they'd been shadowing you for the Society. Nell ran into him by accident. They did not mean you harm," the Thag clarified.

"What? How? Are you trying to tell me that the panther I shot was a human in a cat suit? That's ridiculous!"

Even as Stas said the words, questions nagged at the back of his mind. There was no way it could be true. What the trained thief had just told him had to have a rational explanation. Such things were superstition, he told himself. Backward. He remembered that Kali had tried to tell them much the same thing. That there was some mysticism surrounding the animal Stas had killed. Kali had called it a 'Wobo.' He'd mentioned that his tribe's traditional spears were ineffective against this predator. Stas dismissed it out of hand. It was more of the tribal prince's backward nonsense. Stas considered himself a good Catholic – even though he had been raised in a town without a single place of worship that represented his Roman faith. He'd thought those beliefs reaffirmed when he revealed the spirits conjured by the medicine men of Kali's tribe to be a sham.

Malka fixed the Slav with a determined stare.

"I know that you do not want to believe it. Yet, you have seen it, Stanley. So has Nell. At the Invisible Circus, Balu attacked a tiger to save your friend. He appeared as a panther then, did he not? He rescued her from imprisonment in Pondicherry, as a human. The transformation happened before your eyes, just now. Thanks to Liza's outburst."

"Impossible. It's a trick."

"Stanley, after what you have witnessed in India, that is simply stubbornness. We both know it." The Egypt-born boy clearly did not enjoy having aspects of his own beliefs – his own identity – questioned. It was not a matter of rationality. His reaction was visceral rather than logical.

"Is it? In Africa, the 'mystical spirits' of Kali's tribe turned out to be nothing but drums. It was a total scam. I freed the entire tribe from their ignorance."

"Maybe. Does that mean it is always a fraud? If so, what do you think of my Sect's beliefs? You must admit: you have witnessed things that cannot be explained only by your interpretation of philosophy."

Stas stared back at her. "Why should I believe you? You're a criminal," he snapped.

"That's unfair. Malka's doing what she has to." Henry came to the Thag's defense.

"By your definition, I may be, Stanley. But as I have told you before: if it is so? I am no ordinary one."

The blue-eyed girl sighed. This discussion was getting them nowhere. If Stanley wanted to cling exclusively to his worldview, that was fine. What was important was that they required his assistance. This meant the Thag had to convince her protector to refrain from attacking him. Unfortunately, she reflected, Stanley's attitude was not helping in that regard.

The Thag turned to regard her protector.

"Liza, I regret that this brings back painful memories, but it would appear that what happened was a misunderstanding. He simply wanted to protect his friend. I've seen how much he cares about her. They shared a similar bond to the one that you enjoyed with your partner."

"You heard him. He won't even admit that Tem-vi was a self-aware being."

"Be that as it may, what happened occurred because of coincidence. We all interpret what happens in front of us through the lens of our backgrounds. What happened was tragic. Yet, it occurred because of a clash of worldviews. I am sorry. But you mustn't allow this to escalate. You know, more than most, the strife that can result when an act, meant in defense or desperation, is taken as an affront."

Malka's voice possessed an authority uncommon to one of her years. It was the same tone her own Master had once used to impart knowledge: "We need him. Fighting now will not bring Tem-vi back. It will merely hinder our quest. Adhere to the plan."

Liza continued to glare at Stas.

"Fine," she said eventually. "Doesn't mean I have to like it."

At this point, Stas was able to get a word in edgewise.

"You said you need me? For what?" His head was still reeling. "I know you, Malka. But, who are these two?"

Sighing, Malka explained.

"The one who tried attacking you is my protector. She was assigned to me by the Society. You heard of that back in India. Henry, over there, unwittingly became connected with our group during the course of our journey. He has a good amount of obscure knowledge. It can prove useful, sometimes. I possess knowledge of strategy, deceit and theft. But, mostly I am here because...."

"Because, Prophecy," Henry teased.

"What ... prophecy?" Stas blurted incredulously. His eyes were as wide as saucers.

"Seriously? You mean you don't know either? Perfect." Briefly, Liza outlined what the Prophecy stated.

The Egypt-born youth shook his head.

"Assuming I believe that any of this is true, which I don't, why do you need me?"

"I am sure you have surmised by now, Stanley. I still have the diamond."

Stas nodded.

"Together, we continue to protect it from the Urumi."

"To what end? That Prophecy of yours is unclear. It could mean anything. Someone could have just made the whole thing up and passed it off as fortune-telling...."

"Stanley, you have been to the Invisible Circus. The Prophecy is more than that. We both know it. That is why we require your assistance. We have received information from the Society about what it could mean. We believe it is meant for you to decipher. That means you are destined to become part of our quest."

"*Destined?*" Stas could not keep a disbelieving laugh from escaping his throat as he said the word. "Malka, can you hear yourself? This is crazy. Do you really expect me to believe any of this?"

"When Nell's life was at stake, you were willing to go along with it," Malka pointed out.

"For all the good it did. Nell's.... She's gone. She died in an accident about three months ago."

"What goes around comes around." It was the felinoid snapping under her breath. Now, it was Stas who started toward the felinoid.

"Liza! Put a sock in it. For now," Henry admonished.

"What happened in India were tricks, dressed up by the showmanship of shamans. Nothing more." Something in Stas's voice suggested that as he spoke he was attempting to convince himself, more so than the Thag.

"Then why can't you explain them?" she challenged, directly.

"I don't know." Stas was frustrated. Malka continued:

"You cannot deny that the letters we received from the Society were of assistance...."

Suddenly, a dark apparition appeared in the space between Stas and Malka. Both reeled back.

"Urumi!" Liza yelled. Instantly, she snapped into full combat mode. The black-haired girl leapt towards it, attacking. Colliding, the two crashed into the dorm room's wall. The dark figure offered no resistance. It allowed Liza to pin its substance to the ground. Then, its form resolved.

It was the same blond-haired girl who Malka had fought at the Invisible Circus.

"You again." Liza spat. "What's your business here? Why allow yourself to be captured like this? What kind of a trick are you trying to pull?"

"I can assure you," the black-clothed figure stated quietly, "it is no trick. If I wished to be unrestrained now, I would be. I allow you to hold me for the time being. It is a gesture of goodwill."

"Goodwill? Hah!" the felinoid cracked. "You are capable of no such thing."

"Neither is my kind capable of showing our true form to outsiders. Yet, I do so. I myself do not know how it is possible. If you would let me up, I would offer you a boon."

"Why should we trust you?" It was Malka who asked the question.

"Because, like most Urumi, I was consigned into the service of the Dark Prince unwillingly. But, during our confrontation at the Met...," she described their meeting in the Invisible Circus as if it had been nothing more than a passive-aggressive spat between two operagoers during an interval. "I discovered that – unlike the others – I was able to resist the imperative to do as my Chosen – the one who leads us – commands. As I have said, I should not be able to. Yet, I stand before you now."

She was correct. From what Liza and Malka knew of the Urumi, this should not have been possible. Whether they believed this figure's explanation was another matter entirely.

"Let her up," Malka judged eventually. The Thag figured that if the situation did turn violent, she, Liza, and Stas would be able to handle her. The Egypt-born youth moved to retrieve his rifle, which rested on a shelf above his bed. Liza offered him a glare that could have melted stone. She released the Urumi.

The blue-eyed Slav stood, regarding them. She reached into the folds of her black robes. All present in the room tensed. She moved slowly. Her gestures made clear that they were not intended to be threatening. Her left hand retrieved a small piece of paper. It dropped to the floor in front of Stas.

The half-Slav moved to pick it up. His eyes widened, recognizing it. Back in Madras, he and Mungo had, at first, not gotten off on the right foot. Stas had discovered a small piece of paper, emblazoned with what he now knew to be the emblem of the Society. It had been hidden in the cover of an old book at the library of St. Andrew's school. The paper was filled with a mass of indecipherable lines and symbols. Stas had seen Mungo take notice when he had discovered it. Worried that his rival might attempt to steal it, Stas had hidden the paper in an out-of-the-way location. Upon noticing that his room had been tampered with, Stas assumed Mungo had attempted to purloin the paper. He'd moved to confront the red-haired youth. Shortly after, they had both seen a mysterious dark figure moving away from the school's dormitory at an incredible speed. Realizing that they had a common foe, the two pulled together. Mungo admitted to having found another piece of paper with similarly unintelligible content; it was the article that had just been stolen by the mysterious form, which they'd perceived through the window.

Stas felt sure that this was the paper he now held in his hands. As if by way of confirmation, the warrioress stated: "I took this from your acquaintance's quarters in Madras. I now return it to you. Since that time I have been either shadowing you or Malka."

"Um, excuse me?" Henry cut in. All heads in the room turned from the Urumi to the brown-haired youth. "If you're the one who's been chasing us all this time, why should we trust in anything you give us? I mean, how do we know you didn't tamper with it, or something?"

"He's right. Why the sudden change of attitude?" Liza eyed the blond-haired girl with no small amount of skepticism. "How do we know all this isn't some kind of ploy to lead us into a trap?"

"It is not. I do not know how to decipher the parchment's meaning. Therefore, I could not have tampered with it. I hope that it will aid you in the struggle against the Urumi's designs on the Fragment, which you carry. I now wish to work toward ensuring that such an eventuality does not come to pass."

"Wait. You're defecting?" It was Henry again.

"That's crazy!" Liza snapped. "Urumi can't defect. It's part of what you ... *things* are."

"As I have told you, I recently discovered that I can make my will my own. I do not know how it is possible. As for the veracity of what I have given you? Decipher it. The Betrayers – those you refer to as Arunesh and Zitar – have no doubt sent messages regarding possible courses of action. Compare the two. If the message hidden in what I give you makes sense, heed it. If it does not, feel free to disregard it. But I doubt that you will."

Malka's eyes narrowed. "Why change sides now? Why are you doing this?"

The blond-haired girl shook her head, sadly.

"Because I have a special reason to hate the Urumi and the one chosen to command us."

"Assuming that you are what you appear to be, you still have given me no reason to believe anything you say is true," Stas pointed out, his gun at the ready.

"I am not lying. I do not know how to convince you of that negative. Other than, possibly, to tell you something of how I came to find myself in this situation. It is because of my past that I take these actions now. Yet, I fear that if I tell you these things, it will only make my credibility even more suspect. I simply do not know what else to do." The girl looked squarely at the other Slav in the room, then to each of the others. She began:

"I was born Bozhena Alexevna Lubomirskaya...."

"You're Russian!" Stas spat. His father had rebelled, attempting to free his homeland from the domination of Muscovy. Because of that, the elder Tarkowski had been forced into exile. Stas had been born and grown up in that state of existence.

"No," the black-clad warrioress corrected. She was calm, as if she understood Stas's feelings. "Polish, actually. My mother claimed descent from the Korczak line. My father was heir to the illustrious Lubomirski family's legacy."

"Szlachta," Stas interrupted again. "Even worse. Your kind betrayed our country. You helped foreign powers to seize it. Now you're claiming to betray the ones who command your loyalty, instead. Treachery is in your blood."

"I understand how it could seem that way, Stas. But not all of us were members of that układ." Bozhena's use of the term earned confused looks from everyone in the room but the Egypt-born Slav. His father had used it often. "My parents' marriage was arranged more out of machinations of power and wealth. It had nothing to do with love or like thinking. My mother was a loyal Pole.

"It was my father who sold out to Russia for the promise of a title and a few rubles. That's why my name is the way it is, except for my given name. My mother insisted on that. My father demanded that the Orthodox structure be imposed. That was always the way with him. Everything had to be Russian. He betrayed his own heritage. I hated him for it."

At this revelation, Stas's eyes narrowed with suspicion, which morphed quickly into a cold certainty. Eventually, he accused: "You're Prince Lubomirski's daughter. He tried to kill Nell. And me. Don't believe a word she says!"

Bozhena held up her right hand in a placating gesture. She turned to Malka.

"I can understand why that would make you suspect my pedigree. But has it occurred to you that you are not the only ones to have run afoul of my father's ambition? He drove my mother to suicide. Indirectly, he was the cause of my sister's death. He struck a deal, consigning me into the service of the Urumi, in return for a royal title. Worse, as a dark warrior, I was obliged to work for furtherance of my father's goals. To further his designs on the Fragment, as means to the Dark Prince's ends. I despised my father and was glad for his execution."

"You lie," Malka said coldly.

"Why would you say that?"

"You brought my sister to your father. He shot Antonia," the Thag cursed at her. "If you possessed free will, why? If you tell the truth, you have all but admitted to working with him."

"As I have told, I did no such thing. I was obliged to work for the Dark Prince at that time. I did not know I could do otherwise. But I did nothing to directly aid my father. I did not deliver your sister to him. I had little knowledge of that event until just recently; I learned that your sister was sacrificed out of spite by the Chosen's new aide, a girl once from your region. Not by myself. I am a victim of my father's ambition. As, now, you shall be of yours."

At this, Malka's brow furrowed. The claim the Urumi had just made regarding her sister's kidnapper: it unsettled her. *No,* she thought, *it simply could not be.* If that was a lie, then the Urumi's statement that her father was alive and that he had struck an agreement with the dark warriors could not be, either. Discounting these comments further, the Thag responded.

"I do not believe you. My father is dead."

"No." The blue-eyed Slav stared at the darker-skinned half-breed, sadly. "He lives."

"How do you know?" Malka demanded.

"Because I have had dealings with him."

"Oh, really? How convenient!" Liza drawled sarcastically.

"There's no reason why we should believe any of this." It was Stas. "If anything, what you have said gives me more reason to doubt your sincerity, not less."

"Then I have only one last piece of information to impart. You may find it of interest. However, I fear how you will receive it."

Here, the warrioress turned back to the Africa-born youth. She took a deep breath:

"Your friend. The one you rescued from my father."

"You don't have the right to even discuss Nell." Stas spat at her. "It soils her memory."

"She may still be alive," Bozhena said quietly. Stas simply stared. Clearly, he was interested, in spite of himself. He did not want to believe the person who spoke the revelation. But, if there was *any* chance....

"How would you know?" he asked, guardedly.

"Because in my dealings with your father," the black-dressed girl turned back to Malka, "I was ordered by the Chosen to recruit him. To consign you into our service. Later, I was ordered to command your father to replace Stas's friend with a changeling."

"You did *what*?" Stas yelled, enraged.

"I ordered him to make it appear that she was dead. What her relatives saw die? It was merely a facsimile. What occurs when new recruits undergo the Transmutation before the age of ascent. Your friend was still alive, the last I knew. She'd been sequestered at his manor in Yorkshire. However, I checked before coming here. They are no longer there. I do not know where our cult has ordered them."

Stas threw a few Arabic epithets in her direction. He could not believe what the creature who stood before him now said she'd done to Nell. Even more, she'd had the gall to say so to his face. Her voice had carried a cold, matter-of-fact manner. Yet, Nell's protector was torn. What if there was the slightest possibility? What if the tale this apparition relayed was true? If so, Nell might at least be alive. Albeit, once again in grave danger. Eventually he asked:

"You orchestrated Nell's kidnapping. You expect us to believe that you do not know where she is?"

The descendent of the Korczak line shook her head.

"The Chosen never shares any information with me. All he ever does is issue orders. After my failure in New York, he assigned me to observe you. I no longer control the fate of your friend. I cannot say for certain where she is."

It was Malka who spoke next.

"After all you've done, you expect us to simply take that for granted?"

"Yes. The Chosen never tells me anything. I have ten times his ability. Yet he was selected to lead our Order. Now he has removed me from any mission of import. He took this other newer recruit as his second. I will not tolerate it any longer. Now that I know the ability is mine? I stand ready to assist."

Nobody spoke for a full moment. Finally, Liza said:

"Can you give us any proof – at all – that you're telling the truth?"

"I have told you all that I can. Believe what is written on that which I have given you, or don't. I shall return with more information from my commander, if possible." With that, the black-dressed girl shook her head. Then she was gone.

Stas's mind twisted in confusion. Was Nell still alive? If so, he knew that he had to do anything in his power to find her. Yet, could he trust the information's source? This kind of witch who admitted to being the daughter of a traitorous noble? The same one who admitted doing harm to Nell?

Of course, Stas told himself that changelings were not real. That did not mean he could explain what the black-dressed figure claimed to have done with his closest friend. If there was any chance she was alive, Stas had to act on it.

"Okay, what was up with that?" the boy Malka had called Henry said. Stas jolted out of his reverie.

"Damned if I know," the mysterious black-haired girl said.

"Do you think we can trust her?" Malka asked.

"Nothing she said disagrees with the established facts. But, nothing's proof positive regarding her motives, either," Liza thought out loud.

"Ordinarily, given who she says she is, I wouldn't trust a thing she says," Stas offered. "But if there's any chance that Nell is still alive? I have to try and find her. Even if that means the risk of believing her." Stas set the gun down on his bed as he spoke.

"Look, um ... Stas...." It was Henry, who oddly seemed to have no trouble pronouncing Stas's name. "I get that we only met, um, about ten minutes ago, but I can already tell how important this Nell is to you. If this is all part of a trap, then the, uh, the Urumi might know that. She might just have said it in order to get you to fall for whatever it is they're planning."

Stas sighed. He could not deny the logic that the blue-eyed boy had just laid out in his American accent.

"Henry. This is not about him. Or his little friend," Liza seethed. "This is about the Fragment. So, can we please stick to that and not placating the bruised emotions of a closed-minded killer."

Henry rolled his eyes mildly.

"Um, Liza it's not like he set out to murder Tem-vi. From his perspective...."

"I don't care what he set out to do! Now, we came here to see if he has any information. Let's just get it, so I can get the hell out of here."

Malka allowed the two to say their piece. She shook her head. The Thag suspected that Liza would not have it so easy. If what she surmised were true, Stas had received information relevant to their quest. It was likely they would be required to move on together. That was how things had worked out in Madras. Their enmity was saddening. In other circumstances, the blue-eyed girl suspected, Stanley and Liza would have gotten along.

"Stanley," the Thag began, producing the most recent letter they had received from the Society. "Have you, by chance, received any more of these?"

"Yes," replied Stas. "Two of them. But they don't make sense." He got up and opened up a drawer in his desk, retrieving them.

"I suspect that is because they were intended for us. The Society meant for us to connect, similar to when we received information about the persons we sought in Madras."

"I remember. But I can't see how these will help much." They exchanged the notes that each held.

As soon as Stas read the first line of what Malka had given him, his face flushed with excitement. The Egypt-born youth did not know how. But somehow it was true! He did not want to accommodate the purveyance of these letters within his worldview. Yet, the mysterious leaders of the Society, who ostensibly sent these communiqués, had shown they could be trusted. He jumped up and down a couple of times in elation.

"Nell's alive!" he screamed.

"How do you know?" Malka asked.

"This phrase 'Mała Bint.' The second word's a transliteration. It was what our kidnappers in Egypt used to refer to her. It has to mean Nell. I don't see how it could have made sense to anyone else. You're right. It was meant for me." In his moment of excitement and relief, the half-Slav paused and then exclaimed more to himself than to the others: "Alhamdulillah!"

Malka's trio looked at him oddly.

"I wish I could say the same," Malka said after a moment. She'd taken Stas's meaning, although she did not understand his exclamation. "The first of the two messages you received means little to me. I cannot even read the second."

"That's because it's in Polish," Stas responded. Malka handed it back to him. She gave the other note to Henry and Liza who read its short contents.

"You can translate it," the Thag suggested.

"Uh, yeah, sure I can." The Egypt-born youth pushed a twang of trepidation to the back of his mind as he spoke the words. Stas had always been proud of his heritage; he loved the language connected with it. However, the only person with whom he had been able to converse in Polish, for most all of his life, had been his father. He could understand the message, of course, but he'd had little opportunity to use the language formally. The boy, born of Port Said, stumbled through the translation as he read it back to them.

"'Know. The second well' – no, wait – 'source' or, um 'origin' of the 'force' – no, wait 'power' of the 'Holy Fragment.' 'Lies under the gaze' – I think – maybe it translates more like 'sight – of the Black Goddess. When the appropriate time comes', or 'is arriving. Go to her.'"

"That's exactly what is written there?" Liza snapped, with doubt in her voice.

"Yes. No. Yes. Sort of. Oh, I don't know. It's something like that."

"Can you even speak Polish?" It was Liza again.

"Of course I can." Stas snapped a bit too defensively. "I've never had to translate it before. Sometimes, I don't know exactly what the words mean in English. But generally, the meaning is something like that." It pained Stas to make the next admission. "Mostly, I've only had the opportunity to speak Polish at home with my father."

"Not good enough, damnit!"

"Liza, he's right. Translation can be difficult," the America-born youth weighed in. "It's not like it's a simple one-to-one equivalence."

"Yes, Henry. I know. I speak English, Hindi, Urdu, and *some* Romanian and Hungarian, because I managed to pick those up from my parents. Believe me, I more than get it. But right now I don't care. We need clarity of information. Not educated guessing from someone who speaks kitchen Polish at home with his daddy."

Liza's comments cut at Stas. He had always been proud that he spoke his country's native language, growing up. He didn't enjoy having that questioned. But, what he did not want to admit – even to himself – was that he feared Liza's comments had at least some merit.

"It will have to do, Liza," Malka said. "The general meaning remains clear. Still, without more context, I do not see how we can interpret its meaning, anyway. 'Black Goddess' is clearly a reference to Shakti. However, it does not make sense, literally. According to my faith, if the object which She seeks is already in Her sight we would not need to quest for it. I suppose it could reference a statue of Her black figure. But, there's no actual information as to which one. 'Appropriate time'? That could mean anything."

"My point. We're nowhere. Again."

"I wouldn't say that, Liza," the brown-haired youth puzzled out. "We now know there's a connection between the Fragment and the deity Malka worships – probably a specific statue or figure of it. Likely, this other spring, well or source, is the second mystical object, the location of which was discovered by the Mughals, during the early days of the British East India Company's activity on the subcontinent."

"But what does that actually tell us?"

Henry hunched his shoulders.

Malka pressed forward, questioning.

"Henry, anything regarding the other message?"

"Yes. It also points to Poland or at least east Europe. 'Beware the Noon Witch'? The Noon Witch is a Slavic myth. She was said to replace the children of peasants with sick copies of their children, while taking the real ones. Doesn't that sound familiar?"

"The Urumi," Malka confirmed. "It's a warning against the Urumi."

"So the Urumi are a threat? Really? Wow, that is just *so* helpful. I totally needed a reminder that the dark figures we've been fighting for the past eight months are dangerous. So glad we bothered to come all this way."

"But wait, Liza, it does say Noon Witch. Singular. In light of what just happened, it could be a message not to trust the one that just appeared here."

"Yeah, I get it," she said, glaring at Stas. Henry surmised that her irate manner was merely meant to mask what she must be going through at the moment.

"Stanley, do you have any ideas about what the rest of yours means?" Malka queried.

The green-eyed half-Slav directed his gaze back towards the rest of the parchment. In his elation over the revelation about Nell, and then Liza's derision of his heritage, Stas had neglected to read the rest of it. Some of the phrases made no sense to him, but one section jumped out.

"The 'empty soul of a people self-betrayed,'" he breathed. His father had told him of it. It had been part of the reason that, back in Africa, he had named their hollowed-out tree trunk hut after Poland's former capital. "The soul of Poland. Wawel Castle. The former Polish royal palace, before they moved the capital to Warsaw. This 'noble path'? The old royal road. It leads there. And, Poland. It was partitioned with the cooperation of its nobility. Nell is in Krakow! She must be somewhere along that road!"

"Yeah, I'm way more interested in where we need to take the Fragment," Liza barked at her closest friend's killer.

"What's most important to me is what happens to Nell. I don't know if you can understand that, whatever you are. Anyway, that's what this note is about."

"Maybe it's not. And you just don't know how to interpret the clues correctly!"

"I'm sure of what I've said," Stas said defensively.

"Yeah, what about the rest of it?"

"Could the reference to a patron refer to a patron saint of Poland?" Henry asked.

"I don't know." Stas was forced to admit again, sighing. "I've heard about Krakow all of my life. But, I've never actually been there. None of the other clues make sense to me."

"Have you ever even *been* to Poland?" Liza yelled.

"No!" Stas screamed back in frustration, sitting down. "My father and I aren't even allowed in any of the states that occupy it."

"Amazing! Not only is our Poland expert the one person who I most could have gone the rest of my life without ever seeing again, it also turns out he's never even been there!" The felinoid raised her hands as she yelled. She let them drop back down to her thighs, finishing emphasis of her point.

"Still," Malka countered, "he has done the best he could of learning about it. Even though he was not like the others around himself. Whether he realizes it or not, I understand something of that. In a way, everyone present here does. I trust him. If he needs to go to Krakow to decipher what is on that card, that is exactly what we are going to do."

At this, Stas simultaneously felt relief that he would have some assistance. However, he also felt increasing trepidation. The half-Slav did not want to admit it. However, a deep uncertainty nagged at the back of his mind. All of his life he had been told – indeed, deeply believed – that he was Polish and, within that, European. Then he had come to Europe. Those here had not exactly accepted him. Stas discovered no kinship with its people. Instead, they'd made him feel like a beggar. It now seemed clear that he would have to travel to the country that he had been told he was from his entire life. It was an identity that Stas cherished growing up. Yet, he wondered: how would people there accept him? Would he be rejected as an outsider? If that were the case, what did that make him, but a pariah everywhere he went? The thought induced a deep fear within him.

He'd never belonged, except for when he was with Nell and they were working for their own survival. Now that the prospect of actually going to Poland became real, Stas found that he was afraid. He was terrified that he would be judged unworthy – not accepted as one of the people he had aspired to be for most of his life. In many ways, it would have been easier not to go. To keep the belief that – somewhere – there was a group of people like him alive, in his mind. He would go, but not for his own sense of belonging. He was Stanislaw Tarkowski. Now that he knew Nell was alive, he had to rescue her. That gave him his purpose. The same words he had whispered when he had first risked his life for her, in a desert escape plan, came into his mind: *I'll go. I'll go for Nell.*

"We?" the felinoid yelled. It had taken only seconds for all of this to go through Stas's mind. "Why the hell do you want to go off helping him to find his little friend? We have a much more important mission to complete. Just when I think I've finally gotten you to focus, Malka, you go all half-cocked again. What is it? Some kind of sentimental nostalgia? I'm sick of it."

"Oh come on, Liza, you just don't want to go with him. I mean, do you even have a better idea?" Henry decided to call her out. The felinoid glared at the brown-haired boy. Apparently, she was low on alternate suggestions.

Malka continued. "The Society sent these notes to whom they did for a reason, Liza. Stanley considers it imperative to find Nell. We must determine how to best keep the Fragment from the hands of the Urumi. It is likely the two are connected. We are going together. If we follow one goal, we may achieve the other as well."

"Um, guys?" Henry intervened. "There's one problem. The city Stas just named? We're going to have to cross the Austro-Hungary land border to get there. I read something once. A section of it that connects with Switzerland is called the Tyrol. Part of it is inhabited by Italians who want to break away from the Hapsburgs. This means that border security is likely to be tight. If what Stas says is true?

We're going to have to get across into the neighboring province of Vorarlberg not only without showing passports that most of us don't have. We also need to sneak across completely unseen."

Malka's brow furrowed at this. She was not sure how to surmount the dilemma Henry had just posed.

The Thag started, hearing a gurgle. Jurgen, who had lain forgotten on the floor, was finally coming to. There was a visibly irritated red mark on his neck, where the sash had constricted his airway. He coughed a couple of times. Then, he opened his eyes. It seemed to take the Swiss citizen a few seconds to remember what had happened. When he did, he jumped backwards, towards the wall. Then, the rather formal boy began to skitter on his hands and knees toward the door.

He also began to yelp, his voice still raspy: "Who are you people? Why did you break in here? How dare you attack me! I'm calling for the...," he rasped in French as he reached the door. It was locked. Jurgen realized that the key sat on his desk. It was on the far side of the room. It also registered with him that his roommate was standing nearer to the intruders and that Stanislas was not in a similar state of agitation.

"Stanislas? What is going on here? Who are these criminals? That girl!" he pointed at Malka. "She strangled me."

"I know what she did," Stas stated quietly.

"You know? You mean you are one of them?" The Swiss attempted to hold onto some of his habitual reserve even in this situation.

"No. Yes. Kind of. I don't know, really."

"You let them in to rob the dormitory? Of all the uncivilized, immoral...," the native of Steckborn seemed angrier than Stas had ever seen him.

"No," the Egypt-born youth cut him off.

"Yet, you appear to be fine with their presence?"

"We ... share a common purpose."

"What kind of affairs have you become involved in? I should have known this is how things would turn out. One from the eastern lands, raised in Africa? Of course you would end up sinking to the level of those like yourself." He pointed at the Thag. "What is it? You're going to try and rob this place, aren't you?"

"No." It was Henry, though he spoke French. "Use your overentitled brain. For one, if we were going to rob something, the last place we would have picked is a students' dormitory. For another, if we had, we'd have done it and been gone by now."

Malka approached her erstwhile captive, placing a hand on his shoulder. The Thag had not understood the exchange. "Henry," she asked, as if strangling people into unconsciousness were completely normal. "What was he talking about? He seems agitated."

"Believe me, if you'd understood him, I think you'd have wanted to pull your sash a bit tighter."

Jurgen spoke again.

"He's demanding to know what we're up to," the America-born boy interpreted.

"Tell him none of his goddamn business," Liza snapped.

Henry translated. Stas strongly disliked Jurgen's immediate impulse to imply that the Slavs were somehow of lesser pedigree. At one point in his life, he'd done the same to some of the people around whom he'd lived in the empire. He still could not totally keep himself from doing so now. At this stage, he merely took umbrage at the personal remark related to his heritage. It led him to add, with a hint of condescension in his voice: "They're right, Jurgen. You would be better off not knowing. Now, you need to calm down."

Eventually, Henry and Stas, speaking French, were able to coax Jurgen off the floor and onto his bed. Making introductions and asking Jurgen simple questions, the camp-raised boy had lifted both of his eyebrows upon learning that Stas's roommate was from Steckborn on Lake Constance, and that his father ran a shipping company. Briefly, he cracked a smile. It suggested that Henry's mind had already begun to work on a plan. The discussion switched to a combination of French and English. It turned out that the Jurgen's command of English was such that he could understand much of what was said, but he preferred to respond in French.

Stas let him know he had reason to believe that Nell was still alive and that she was in the Hapsburg-controlled territory of what had formerly been Poland. However, the native of Port Said had refused to divulge how he had come by this information. He did not mention anything about the diamond that the darker-skinned girl carried. It was fortunate, Malka reflected, that this local person had not regained consciousness until after the departure of the Shadow Warrior and discussion of their next steps. That was almost the only thing that had gone smoothly about this meeting.

Jurgen had demanded of his roommate how he had allowed himself to become involved with "such people." It was as if he thought the Africa-born youth's decision flowed from some sub-Swiss character flaw, rather than the circumstances in which the Slav found himself.

Stas had merely glared at Jurgen, a mysterious gleam in his eye.

"Do you really want to find out?"

The Swiss youth sat for a moment in frustrated silence.

Eventually, Henry spoke.

"Really, Jurgen, the less you know the better … for your own safety." The boy paused to let the implications of that comment sink in. It was true; the less Jurgen knew about their plans, the less likely the Urumi would target him. However, if the suit-dressed boy chose to take the comment as an implied threat, so much the better. "The only reason why we've told you as much as we have? It's because we require your assistance. You see, we need to get into Austria without, shall we say, the knowledge of the Emperor's armed forces." Henry was clearly enjoying himself.

"You mean…," Jurgen began, processing the implications of that last statement. "You want me to help you sneak illegally across the border?" The formal Swiss German stood abruptly. "How dare you even suggest that an upstanding, civilized person could be capable of an act only the likes of you would commit? I will never condone…."

Henry interrupted, deciding to translate this statement for the non-French speakers in the room. Some laughed harshly.

Malka put a firm hand on Jurgen's shoulders, shoving him roughly back onto his bed. She stared at him coldly with her blue eyes.

"What exactly do you mean by 'the likes of you,' Jurgen?" Stas asked. There was a coldness to his voice that the Egypt-born boy was unaccustomed to hearing from himself.

"You. A backward, oriental Slav who's spent so much time around the primitives that he's become one of them. An American hick. A violent gypsy girl, and a peasant from the Romanian hinterlands." His description of Liza was on the basis of the last name Henry had given Jurgen. Malka had simply been introduced as Malka.

"That is who I mean. It truly disgusts me, Stanislas, that you would fall in with the likes of them. I should have seen it. You aspire to be one of us, but you will always be like one of them. It is in your constitution."

"When did I *ever* say that I aspire to be like you?" Stas glowered. He recognized the arrogance: two years ago, he'd held himself, unquestioningly, in the same regard. Even now, there were aspects of his own identity that he refused to concede. "I am proud of my past. My heritage. I have wanted to tell you this from the first time we met. You have no right to say that I am less worthy than you, simply because of who you think you are," the half-Slav finished.

"Yeah, really? Who'd have thunk it!" Liza snapped at Stas. The Egypt-born boy's brow furrowed at this. It seemed intuitive, what he just had said to Jurgen. Of course it should apply to his roommate, but could it apply to him as well? He did not want to countenance the thought. It was yet another nag at the back of his mind.

Jurgen shook his head, as if in wounded pity.

"Then, I wonder, Stanislas. What does it say that you don't aspire to be like us? You cannot even see that our way is better? Are you so blind to that fact that you consort with those who think that a member of the upper echelons of Swiss society would stoop to your level? Without even a thought to morality, as befits a member of our exceptional state? I must confess: I do not know what is more sad or revolting in all of this."

Henry favored the Swiss youth with an epic eye roll, combined with a loud exhalation.

Silence hung in the air for a beat. Then Stas found that he was speaking. Two years ago, he could not have imagined himself saying such a thing, especially in an attempt to force another into an illegal venture. True, he had attempted to play upon the fears of his Sudanese kidnappers. Yet, that had been in an attempt to dissuade them from their own course of extralegal action. Now, he was surprised by how little different this felt. In both cases, he'd simply done what he'd thought necessary.

"But, for how much longer, Jurgen?" His mouth formed the words.

"What?"

"For how much longer? I know about your father's shipping troubles. For all your pomposity, one turn of fate and you're no better." Quickly, Stas told the rest of the party of Jurgen's misfortune. Henry and Malka smiled. Jurgen simply appeared incensed, but did not possess the English skills to suddenly interrupt.

"How dare you!" he'd thundered in French as Henry translated. "You share my private matters with these upbraids? Without my consent? Besides, how could lowlifes like you hope to do anything? There's no way the likes of you could have enough funding to resolve my family's troubles." His righteousness gathered resolve. "It's why your people remain where they are, while my people create civilization. So even if you have the means to resolve my family's difficulties, I am repulsed by the imp...," the Swiss trailed off in mid-sentence.

Malka and Henry were staring at him with strange smiles on their faces. Eventually, it was Henry who broke the news:

"Um, yeah," he started, drawling out the second word. "Here's the thing...."

Seventeen

The horse's hooves clopped over the dusty path that led along the lake's southwestern shore. The animal was attached to a plain wooden cart. Jurgen was driving. Stas rode next to him. Liza, Henry, and Malka sat crouched in the back. They were on top of what appeared to be a large wooden plank.

The morning after their hectic meeting in the Fribourg dorm room, they had set out for the Swiss German's home in Steckborn. Their journey had taken them first, by train, to Zurich. Then, with Malka paying for yet another new set of first-class tickets, the group moved on to the lakeside community.

For the rest of that night in the dorm of St. Nicholas School, Henry, Malka, and Liza had recounted the unlikely story of how they had successfully heisted a bank. At many points, the Swiss youth had blanched, turned to Henry and asked him to confirm what had been said in German or French. He had not believed he'd understood correctly. Henry encouraged Jurgen to trust his English. After some further meowing – figuratively from the upper-class Swiss German over what he would be required to do in order to secure funds from the mysterious grouping that had invaded his quarters in the middle of the night, and literally from Liza over the general tenor of Jurgen's objections – the native of Steckborn had eventually agreed to assist the four across the border, unnoticed, in return for enough funds to help rescue his family's business.

In Steckborn, they had spent two days as guests at the Fischer family's home. It was located near the city center. The large row house presented a facade built in the half-timbered Swiss style. No shortage of servants kept it up. The group had met Jurgen's parents. They'd seemed just as formal as their son. The suit-dressed youth had introduced Stas and Henry as friends he'd met at school. Malka and Liza were explained as friends, in turn, of theirs; the two girls could not speak French and, as such, could not have been classmates. Generally, they had all been made to feel welcome.

Yet, two things concerned Stas during that time. The fact that they were about to deliberately sneak across international borders, surprisingly, did not trouble him. A greater source of trepidation came from the fact that doing so would set him on a course towards finally entering Poland: would he find kinship with the people in that land, which he had never visited? Or – he did not allow himself to consciously think this thought – would he be rejected as a foreigner?

More practically, once in Steckborn and not under the immediate supervision of Jurgen, the four had resumed discussion regarding the mysterious piece of paper given to them by the apparently errant Urumi. Stas had wanted to ignore it. Anything given to them by one who had assisted in Nell's kidnapping must be false, or a forgery, he reasoned. At any rate, it could not have any value. Malka had initially seconded him. The Thag cited similar reasons of personal history.

The felinoid and Henry dissented. They maintained that something which might be less than truthful could still serve as an important source of information. It could help tell them what was going on in the minds of their foes. Malka quickly relented, once she was brought to confront this line of reasoning.

"We have to find out what its message means," Malka opined. "Even if there is only a chance that it is genuine."

"Very well."

It was agreed. They had asked to use the house's kitchens. There, as Malka and Stas had seen Mungo's father do in Madras, they finely chopped the head of one cabbage. The two boiled the strips and poured the resulting broth on the paper. They waited for the hidden message to appear.

Nothing happened.

"This worked in the past. Where is the message?" the Thag had asked, more in confusion than of anyone in particular.

Liza chose to treat said inquiry as more than rhetorical.

"It means that either you two don't know what you're doing, that the Urumi gave us a fake, or that this message only becomes apparent after you subject it to a different process that we don't know about."

"Great. No, really. Just great," said Henry.

With nothing left to do, the group had said goodbye to Jurgen's parents, who had been told that they would be traveling on to Vienna and then to Krakow. Jurgen volunteered to drive them across the border personally, saying to his elders that he would take them to the nearest town that lay within the territory of the Hapsburg Empire. This, of course, was not technically a complete fabrication. Stas had felt more than a twinge of guilt at having to lie, even if he wasn't the one who personally did so. Protecting Nell and getting her to safety had once been served by what he had always thought were the habits of an upstanding Polish gentleman; the opposite was now undeniably called for in this situation.

Jurgen slowed the wagon. He steered the horse off the main path. They bounced over an empty field. The last rays of sunlight could barely be seen over the western horizon. Presently, the conveyance reached the water's edge.

Stas was surprised. Of course, border patrols were too frequent to risk attempting a crossing by land. Yet, especially knowing that Jurgen's father ran a shipping company, the Slav had expected to see a boat of some sort. There was nothing present but water, lapping lazily against a thin shoreline.

"Jurgen, where is the boat you promised?" the native of Port Said asked pointedly, before they had even gotten out of the wagon. The other three reacted with equally suspicious and confused looks. Henry translated Stas's question from French for the other two present.

"I never promised any boat," the Swiss replied begrudgingly. "If you'd think about it, using a ship to get across would not work." Clearly, the Swiss-born youth was not happy about being in this situation. "Once you'd docked it would be noticed. You'd have to go through border and customs control." Jurgen spoke as if annoyed that his erstwhile roommate had not thought of this. "Also, my father would ask what was going on. It would be out of the ordinary if I asked him to crew an entire ship just for you … people."

"You said that you would get us…," Stas countered angrily.

"I said that I would. I do not say those kinds of things if I cannot," Jurgen replied as he got down from the wagon. The one who had grown up on the lake motioned for the remaining four to do the same.

"Help me with this," the Swiss German ordered, still in charge by entitlement.

Hesitantly, the other four helped remove the long polished board from the wagon bed. They carried it to the water. It was heavy, but being made of wood, it floated.

"When I was younger, I used to enjoy paddling around the waters of Steckborn on this piece of equipment," Jurgen explained.

"You want us to float across the border on *that?*" It was Liza who hissed the question. "*At night?*"

"It is the best I can do. But I think it will work."

"If you really expect us to give you half a million francs for a damn *raft ride,* then you're out of your Helvetic mind!"

"I will not be coming with you." Jurgen picked up on the felinoid's implication. "I would not be able to explain my presence in Austria on my return without the presence of a proper entry stamp in my passport. And yes, I do. I may not know the entirety of what you people are doing. However, I can discern that you are desperate. As am I."

"Oh, God damn you. Very well."

"All you need to do is go out into the lake until the shore is barely visible. Float across until you see a marshland begin and end. From there, you can kick back to shore.

"*Kick!*" Stas yelled. "Jurgen, for all your pontification about honesty. You promised to…."

"Get you across the border unnoticed. That is what I am doing."

The Slav harrumphed.

"If you hold onto the edges of the waterboard, with your belongings on top, your profile should be low enough in the water that you'll appear like waves to any watch patrol."

Henry had been translating the exchange. He was the first to notice an increasing fear in the Thag's eyes.

"Um, guys, wait," he said, turning to the camp-raised girl.

"Malka? Can you swim?" he asked.

"I'm afraid that there was no body of water in my camp. I never had an opportunity to learn," she said quietly.

"We can't do this." Henry's response was immediate. Decisive.

"This may be our only option. If it's what we have to do in order to rescue Nell, then I have to. And, apparently you need me for some reason. We're going." Stas's opposition was voiced in a tone equally unlikely to yield.

"We can't. I won't allow it," the brown-haired boy was adamant.

Malka moved to place a calming hand on his shoulder. "I am grateful for your concern, but if this is what I must do, I will."

"If she's fine with it, so am I," Liza decided. "All she has to do is hang on to a side of our half-million franc piece of wood and not let go."

"One of us will be there to grab a hold of her, if she slips," Stas reassured.

"All right." Henry sighed. They moved towards the wooden raft. First they piled their belongings on its top.

Then, Malka handed one of the three saddlebags full of currency over to Jurgen.

"I'm told that this sum should cover the amount you require," the camp-raised girl explained.

"Don't give it to your father or a bank immediately." the felinoid ordered. "They might guess where you got it. At any rate, it sounds like you have time before your company craters. If people ask questions, your name will have to come up pretty quickly. And I'm guessing that the cops aren't going to be too happy with what you've done here." The felinoid called to Jurgen over her shoulder as she got into the water. Liza pushed the raft a bit out from shore, so that the others could get in position.

"I understand," Jurgen responded.

They took hold of the raft. Henry and Liza were on one side; she had not wanted to be next to Stas. Malka and the Egypt-raised youth were on the other. They pushed off the lake's soft bottom. Soon it fell away from under them, completely. Three out of the four kicked softly. The group followed the course, which their facilitator had indicated.

As they did so, Henry turned to Liza and whispered.

"Liza, do you like swimming?" It seemed like a ridiculous question, asked in the midst of a serious time. They could still be caught. The simple fact that Stas's roommate had sent them this far did not preclude that he could still double-cross them by alerting the border police.

"Later, Henry," the felinoid hissed back.

"I only ask because most cats hate water. Is it the same with felinoids? I mean you could have just walked across the border. You know, in your other form."

"We do. Unless, of course, one of us manifests as a Turkish Van, or some other rare, water-loving breed. But right now I feel better being here to protect you people, in case this all goes south. And, as long as I'm going to be here, I may as well provide a little added propulsion. One of us has never been waterproofed, if you know what I mean."

"Okay, uh, thanks, I guess."

"You have no idea, Henry," the black-haired girl replied, making a show of her response. "No idea...."

Meanwhile, on the other side of the raft, the Thag and Stas were having their own discussion. Noticing the expression on Malka's face, he'd asked, "Are you nervous?"

"Yes. A bit."

"You shouldn't worry too much. There are three other people here to catch you if your hands should slip."

"My hands never slip," the Thags' leader responded, as if taking the remark as a challenge to her martial abilities. Then, her expression softened, slightly. "It is not that. I'm just hanging here. I am worried about what is going to happen when we get to our destination. I wonder if, somehow, I will not be able to complete the quest with which I have been entrusted, because I am not truly a Thag. It isn't the water, really. I've ceased fearing death, Stanley. What I fear is failure. If that comes, I'm afraid of being left truly alone with that knowledge. Or, worse than that, I'm afraid I'd become an object of pity."

The Slav nodded.

"I think I understand," he said. How many times had he risked his life for Nell? It gave him a clear purpose. But, the costs of failure were too enormous to even contemplate. "No matter what I may think of what your people do, I am forced to wonder if, now, they did not have a reason for doing so. One that was honorable – for them – in its own way. Whatever my own beliefs, being the last of them cannot be easy. You and I are placed uniquely to understand that. But the knowledge that you have failed to fulfill the duty you hold most important…," the Slav paused for a second. "Well, that just scares me to the core."

The Thag regarded her fellow traveler for a moment. She noticed that her words had brought a worried expression to his face.

"Stanley? Do *you* know how to swim?" the blue-eyed girl asked.

"Do I know?" he responded. "I'm swimming right now."

"You are the one who looks worried."

He paused for a moment, sighing.

"It's not the water or even this crazy border crossing. It's just…."

"What?" the Thag pressed.

"Well, you know what it's like to feel isolated. Like you are the outsider, no matter where you are."

"Yes, of course. We had that discussion back in Madras."

"I know. But what if you had been told all of your life that somewhere out there, there was a group of people who were like you?"

The Thag considered this for a moment. Unlike Stas, this had never been the case with her. She supposed that she had believed, vaguely, that those outside the camp were like her, before she had begun her trainings with Husain. But that – the blue-eyed girl supposed – was different from having been told specifically that such a group existed, and that you were a member of it.

"I-I am uncertain."

"Would you want to go to them?"

"I suppose?" the Thag responded, though she sounded unsure. To Malka, the dilemma was understandable.

"And once you were on the way there?"

"Excited?"

"That's what I always thought. See, Malka, I was told that I was Polish all my life, but I have not been accepted as one of the people on this continent. That's been because of my upbringing, as much as because of my bloodlines. I think I'd always been aware of the fact on some level. But, as long as I lived in the empire it was never a problem because…"

"You could use it as an excuse. A reason to blame, for the lack of simple solutions in existence," Malka finished for him. She thought of her own experiences, how she'd used being forbidden from leaving her camp as an excuse. She had pointed to it to explain why those in the Thag's village treated her differently. She also knew how quickly her world had unraveled, once that out was no longer there.

"Sadly." Stas offered only one word of reply.

"That is what you fear. Am I correct? That they would reject you as one of them."

"Yes." For all of his past bluster, the Egypt-born youth's voice was small.

"If I were in your situation now, I believe that I would be … how did you put it? Scared to the core?"

The Slav nodded.

"What can we do about it?"

Malka sighed. "Try to fulfill our purpose for as long as we have one. In the name of whatever – whomever – we think we may belong with."

"What if 'where we belong' turns out to be nowhere?"

"I do not know." Malka again offered a pained look. "I have found it best, lately, not to think much of the future."

Now it was Stas's turn to sigh deeply as he kicked. After a moment, he said:

"What are the odds that we would come from such different backgrounds – different sets of beliefs and values – and still have so much in common? All my life I'd been taught that was impossible."

Malka smiled briefly. She shook her head.

"It is unlikely. Those differences were difficult for both of us to overcome, at first. And, what you say? It's true of them as well." The Thag cocked her head to the other side of the raft. "You know something of Henry's background in the mining camp. If you'd look past your suspicion of Liza's nature, I believe you'd find much in common as well. For what it's worth, I am glad to have you along on this most dangerous of quests, with high odds of failure and a very uncertain outcome."

Stas had to keep himself from laughing out loud. He'd had to compose himself before replying in a whisper.

"I'm still a bit doubtful of her nature. But, doing things like this to save Nell? I'm in no position to act high and mighty anymore. It's she who won't talk to me. How could I have known that what looked like a panther was supposedly...," he trailed off, nostalgic not for the specific memory itself but for the mindset of those times. "The world looked so simple then." The Slav eventually exhaled, in longing.

The Thag nodded in understanding. "It is what I have tried to tell her."

Both sighed.

"I guess pushing forward with you guys is all I have. For as long as that lasts."

Again, Malka frowned; a sad look crossed her expression.

"I am afraid that it is all any of us can do. I would counsel you to accept it. Except, I can't quite seem to take my own advice," she said softly.

The landing zone that Jurgen had mentioned was coming into view. A marshland ended before the lake's shore jutted outward, suddenly. A short, muddy beach began.

Then they noticed it: the shapes of uniformed figures holding lanterns. Some of them held canines on leashes. The dogs could be heard barking loudly as the group came closer.

"That hypocritical little son of a...," the felinoid vented coldly.

"He double-crossed us," Stas said. Steel in his voice, he cut the end of the green-eyed girl's rant short.

The four stared at the assembled mass of law enforcement officers who observed the dark waters of Lake Constance. Their attack dogs barked as they picked up the invaders' scent. Once freedom – hope – had lain in waiting at this lake's farther shore.

Now, only failure beckoned.

Eighteen

Nell had no idea where she was. She'd spent about a week on a ship, locked in a room. Eventually, they'd arrived at the port city that she assumed was the one she had heard the boat's porter mention upon her forced embarkation, in a metropolis she'd not recognized as London. Danzig, he'd called the ship's destination. Nell had studied geography. Miss Oliver had been her main tutor, of course, back in Egypt. But, Stas had always been there to help with homework. He had not been the most patient of teachers, but the girl had always admired the intelligence of the boy, who had been her next-door neighbor for as long as she could remember.

The girl had studied the atlas that had been on the shelves in her gilded Yorkshire cage. Yet, she had not concentrated on the northern parts of Europe. Nell could not place the city in which the ship had arrived. *Somewhere in Germany?* She'd contemplated on the voyage. But, she'd looked at the maps of parts of that country too. The name had never come up.

Directly after the vessel's arrival, the high-ranking British officer, who she'd come to think of as 'her captor,' had muscled her directly into a carriage. It had moved along an industrial waterfront before turning right, onto a longish cobblestoned road. The ornate buildings on either side of it took her breath way. *What is this place?* she'd wondered again, at the time.

To her perception, the ride had taken days, maybe longer. However, the light brown-haired girl was not sure. She suspected that the man in the elaborate British military uniform had kept her at least partially sedated – maybe something in the rations of food he'd provided. All of the time, the aging brown-haired man sat diagonally across from her in the wagon. She'd tried to ask questions, whenever she was able. He never replied.

Nell had considered a hunger strike. But no, she'd concluded. Stas would find her – somehow. When that day came, she would need to be ready to escape with him. That meant she needed to keep her strength up, no matter what might be in the food.

Then, another transfer occurred. It was dark. Nell was drowsy. Her eyes could not quite discern the surroundings. She was ushered through a small door in what appeared to be a brick and stone building.

Now, she sat strapped to the post of a dark wood bed frame in a rococo-style chamber. If Nell craned her neck, she could partially see out of the room's small, cast-iron grated window. Beyond it, a green space was backed on one side by what looked to be a building of some sort. Yet, she could not see enough of it to determine exactly what type of structure it was.

Except for being allowed up – under the direct supervision of her captor – that was how Nell had spent the intervening days. She did not know exactly how many. Sitting there, chained to the bed, doubt began to creep into her mind. For the first time, she truly considered her captor's words:

What if everyone does think I'm dead? Then, how can Stas know to come for me? Even if he tries, it's pointless. How can he know where to find me? I don't even know where I am.

No. Another part of her mind had countered, with faith. *Stas can do anything. You've always believed that!*

He's only human. He cannot find what he thinks is no longer there to find, the other voice in her mind replied with uncharacteristic force, cynicism. *You cannot even help yourself. You tried. You failed. Now you sit here chained to this bed. You can't always rely on Stas to save you. If you don't know where you are, how can he? Even if he did, it is no* guarantee *that he could save you.*

I think it's time to consider: This may be it*, Nell.* The other voice conceded. *It has been months.*

She craned her neck to look through the grated shut window at the people who passed in front of the green sliver that was all her eyes could see. An emotion overcame her. It was not fear. Nell had felt that before, on multiple occasions. This went far beyond that: the realization that this could be 'it.' Stas's friend did not cry. She began to mutter a prayer to anyone, any thing or deity that would hear:

"Please, let it not be. Please, let it not be. *Please*, let it not be...."

She had been kidnapped twice before. For the first time, Nell had to consider one stark possibility:

Rescue will never come.

Have faith

Our quest continues in

Keepers of the Stone

Book Three:

Homecoming

About the Author

Interested in his Slavic roots from an early age, Andrew Anzur Clement spent his teenage years devouring a good number of sci-fi/fantasy books, during which time he first encountered the characters of Stas and Nell. While working and studying in Central Eastern Europe and South Asia, he discovered an insatiable wanderlust that has taken him to many fascinating places across the globe. Now in his mid-twenties, the native of Southern California lives on the far side of the Atlantic. He can sometimes be found slinking through the nosebleed sections of Europe's opera houses, searching for inspiration.

www.ingramcontent.com/pod-product-compliance
Lightning Source LLC
Chambersburg PA
CBHW032013240626
47153CB00003B/1242